T0297117

CAMBRIDGE LIBRARY COLLECTION

Books of enduring scholarly value

Darwin

Two hundred years after his birth and 150 years after the publication of 'On the Origin of Species', Charles Darwin and his theories are still the focus of worldwide attention. This series offers not only works by Darwin, but also the writings of his mentors in Cambridge and elsewhere, and a survey of the impassioned scientific, philosophical and theological debates sparked by his 'dangerous idea'.

On the Connexion of the Physical Sciences

Mary Somerville (1780–1872) would have been a remarkable woman in any age, but as an acknowledged leading mathematician and astronomer at a time when the education of most women was extremely restricted, her achievement was extraordinary. Laplace famously told her that 'There have been only three women who have understood me. These are yourself, Mrs Somerville, Caroline Herschel and a Mrs Greig of whom I know nothing.' Mary Somerville was in fact Mrs Greig. After (as she herself put it) translating Laplace's work 'from algebra into common language', she wrote On the Connexion of the Physical Sciences, published in 1834.

On the Connexion of the Physical Sciences

MARY SOMERVILLE

CAMBRIDGE UNIVERSITY PRESS

Cambridge New York Melbourne Madrid Cape Town Singapore São Paolo Delhi

Published in the United States of America by Cambridge University Press, New York

www.cambridge.org
Information on this title: www.cambridge.org/9781108005197

This edition first published 1834
This digitally printed version 2009

ISBN 978-1-108-00519-7

ON

THE CONNEXION

OF

THE PHYSICAL SCIENCES.

BY

MRS. SOMERVILLE.

LONDON:

JOHN MURRAY, ALBEMARLE STREET.

MDCCCXXXIV.

To the Queen.

—o—

MADAM,

IF I HAVE SUCCEEDED IN MY ENDEAVOUR TO MAKE THE LAWS BY WHICH THE MATERIAL WORLD IS GOVERNED MORE FAMILIAR TO MY COUNTRYWOMEN, I SHALL HAVE THE GRATIFICATION OF THINKING, THAT THE GRACIOUS PERMISSION TO DEDICATE MY BOOK TO YOUR MAJESTY HAS NOT BEEN MISPLACED.

I AM,

WITH THE GREATEST RESPECT,

YOUR MAJESTY'S

OBEDIENT AND HUMBLE SERVANT,

MARY SOMERVILLE.

Royal Hospital, Chelsea,
1 *Jan.* 1834.

PREFACE.

The progress of modern science, especially within the last five years, has been remarkable for a tendency to simplify the laws of nature, and to unite detached branches by general principles. In some cases identity has been proved where there appeared to be nothing in common, as in the electric and magnetic influences; in others, as that of light and heat, such analogies have been pointed out as to justify the expectation, that they will ultimately be referred to the same agent: and in all there exists such a bond of union, that proficiency cannot be attained in any one without a knowledge of others.

Although well aware that a far more extensive illustration of these views might have been given, the author hopes that enough has been done to show the connexion of the physical sciences.

ERRATA.

Page

46, line 5, for "13th," read "30th."

56, line 6 from bottom, for "discrepancies," read "discrepances."

57, for lines 3d and 4th, read "gives $\frac{1}{298 \cdot 33}$ for the compression deduced from arcs of the meridian."

59, *Note.*—The effect of local attraction on the pendulum is so great, that it has rendered the experiments made with that instrument for the purpose of ascertaining the compression of the earth very uncertain. Mr. Baily, President of the Astronomical Society, has devoted much attention to the investigation of this subject. He finds that the experiments of Captain Foster, whose early loss is so justly lamented, give a compression of $\frac{1}{289 \cdot 48}$; those of Captain Sabine give $\frac{1}{288 \cdot 40}$; the mean of the French and Russian experiments give $\frac{1}{267 \cdot 23}$; from the mean of the whole Mr. Baily deduces the compression to be $\frac{1}{285 \cdot 26}$; but even this is not conclusive.

64, line 7, for "92246700," read "95296400."—Line 8, for "ninety-two," read "ninety-five."

Note.—If the computation be made with the more accurate parallax $8'' \cdot 5776$, the sun's distance is 95070500 miles.

67, line 8, the quantity $\frac{1}{1053 \cdot 924}$, representing the compression of Jupiter, was not deduced from Encke's comet, but from the perturbations of Juno.

Note.—Professor Airy has recently determined the most accurate estimation of the value of the mass of Jupiter to be $\frac{1}{1048 \cdot 69}$ deduced from the elongation of the fourth satellite: he has also found that the mass of the whole Jovial system is $\frac{1}{1048 \cdot 70}$, showing how small a proportion the mass of the satellites bears to that of the planet.

68, line 8, for "886860," read "886952."

88, lines 7 and 8 from bottom, for "radius," read "diameter."

99, line 7, for "poles," read "pole."

128, lines 1 and 2 from bottom, for "volume," and "volumes," read "atom" and "atoms."

129, lines 1 and 3, for "volumes" and "volume," read "atoms" and "atom."

132, line 9, for "freezing," read "zero."

144, line 11, for "1090," read "1123."

220, line 3 from bottom, for "rays," read "images."

SECTION I.

All the knowledge we possess of external objects is founded upon experience, which furnishes facts; and the comparison of these facts establishes relations, from which induction, the intuitive belief that like causes will produce like effects, leads to general laws. Thus, experience teaches that bodies fall at the surface of the earth with an accelerated velocity, and with a force proportional to their masses. By comparison, Newton proved that the force which occasions the fall of bodies at the earth's surface, is identical with that which retains the moon in her orbit; and induction led him to conclude that, as the moon is kept in her orbit by the attraction of the earth, so the planets might be retained in their orbits by the attraction of the sun. By such steps he was led to the discovery of one of those powers with which the Creator has ordained that matter should reciprocally act upon matter.

Physical astronomy is the science which compares and identifies the laws of motion observed on earth with the motions that take place in the heavens; and which traces, by an uninterrupted chain of deduction from the great principle that governs the universe, the revolutions and rotations of the planets, and the oscillations of the fluids at their surfaces; and which estimates the changes the system has hitherto undergone, or may hereafter experience—changes which require millions of years for their accomplishment.

The accumulated efforts of astronomers, from the earliest dawn of civilization, have been necessary to establish the mechanical theory of astronomy. The courses of the planets have been observed for ages with a degree of perseverance that is astonishing, if we consider the imperfection and even the want of instruments. The real motions of the earth have been separated from the apparent motions of the planets; the laws of the planetary revolutions have been discovered; and the discovery of these laws has led to the knowledge of the gravitation of matter. On the other hand, descending from the principle of gravitation, every motion in the solar system has been so completely explained, that the account of no astronomical phenomenon can now be transmitted to posterity of which the laws have not been determined.

Science, regarded as the pursuit of truth, which can only be attained by patient and unprejudiced investigation, wherein nothing is too great to be attempted, nothing so minute as to be justly disregarded, must ever afford occupation of consummate interest and subject of elevated meditation. The contemplation of the works of creation elevates the mind to the admiration of whatever is great and noble; accomplishing the object of all study,—which, in the elegant language of Sir James Mackintosh, 'is to inspire the love of truth, of wisdom, of beauty, especially of goodness, the highest beauty, and of that supreme and eternal Mind, which contains all truth and wisdom, all beauty and goodness. By the love or delightful contemplation and pursuit of these transcendent aims, for their own sake only, the mind of man is raised from low and perishable objects, and prepared for those high destinies which are appointed for all those who are capable of them.'

The heavens afford the most sublime subject of study which can be derived from science. The magnitude and splendour of the objects, the inconceivable rapidity with which they move, and the enormous distances between them, impress the mind with some notion of the energy that maintains them in their motions with a durability to which we can see no limit. Equally conspicuous is the goodness of the great First Cause, in having endowed man with faculties by which he can not only appreciate the magnificence of His works, but trace, with precision, the operation of his laws; use the globe he inhabits as a base wherewith to measure the magnitude and distance of the sun and planets, and make the diameter of the earth's orbit the first step of a scale by which he may ascend to the starry firmament. Such pursuits, while they ennoble the mind, at the same time inculcate humility, by showing that there is a barrier which no energy, mental or physical, can ever enable us to pass: that however profoundly we may penetrate the depths of space, there still remain innumerable systems, compared with which those apparently so vast must dwindle into insignificance, or even become invisible; and that not only man, but the globe he inhabits,—nay, the whole system of which it forms so small a part,—might be annihilated, and its extinction be unperceived in the immensity of creation.

Although it must be acknowledged that a complete acquaintance with physical astronomy can be attained by those only who are well versed in the higher branches of mathematical and mechanical science, and that they alone can appreciate the extreme beauty of the results, and of the means by which these results are obtained, it is nevertheless true that a sufficient skill in analysis to follow the general outline,—to see the mutual dependence of the different parts of the system, and to comprehend by what means some of the most extraordinary conclusions have been arrived at,—is within the reach of many who shrink from the task, appalled by difficulties, which, perhaps, are not more formidable than those incident to the study of the elements of every branch of knowledge; and who possibly overrate them from disregarding the distinction between the degree of mathematical acquirement necessary for making discoveries, and that which is requisite for understanding what others have done. That the study of mathematics, and their application to astronomy, are full of interest, will be allowed by all who have devoted their time and attention to these pursuits; and they only can estimate the delight of arriving at the truths they disclose, whether it be in the discovery of a world or of a new property of numbers.

SECTION II.

It has been proved by Newton, that a particle of matter, placed without the surface of a hollow sphere, is attracted by it in the same manner as if the mass of the hollow sphere, or the whole matter it contains, were collected in its centre. The same is, therefore, true of a solid sphere, which may be supposed to consist of an infinite number of concentric hollow spheres. This, however, is not the case with a spheroid; but the celestial bodies are so nearly spherical, and at such remote distances from one another, that they attract and are attracted as if each were a dense point situate in its centre of gravity,—a circumstance which greatly facilitates the investigation of their motions.

The attraction of the earth on bodies at its surface in that latitude the square of whose sine is $\frac{1}{3}$, is the same as if it were a sphere; and experience shows that bodies there fall through 16·0697 feet in a second. The mean distance of the moon from the earth is about sixty times the mean radius of the earth. When the number 16·0697 is diminished in the ratio of 1 to 3600, which is the square of the moon's distance from the earth's centre, it is found to be exactly the space the moon would fall through in the first second of her descent to the earth, were she not prevented by the centrifugal force arising from the velocity with which she moves in her orbit; so that the moon is retained in her orbit by a force having the same origin, and regulated by the same law, with that which causes a stone to fall at the earth's surface. The earth may, therefore, be regarded as the centre of a force which extends to the moon; and, as experience shows that the action and re-action of matter are equal and contrary, the moon must attract the earth with an equal and contrary force.

Newton proved that a body projected in space will move in a conic section, if it be attracted by a force directed towards a fixed point, and having an intensity inversely as the square of the distance; but that any deviation from that law will cause it to move in a curve of a different nature. Kepler ascertained, by direct observation, that the planets describe ellipses round the sun; and later observations show that comets also move in conic sections: it consequently follows that the sun attracts all the planets and comets inversely as the square of their distances from his centre; the sun, therefore, is the centre of a force extending indefinitely in space, and including all the bodies of the system in its action.

Kepler also deduced from observation, that the squares of the periodic times of the planets, or the times of their revolutions round the sun, are proportional to the cubes of their mean distances from his centre: whence it follows that the intensity of gravitation of all the bodies towards the sun is the same at equal distances; consequently gravitation is proportional to the masses, for, if the planets and comets were at equal distances from the sun, and left to the effects of gravity, they would arrive at his surface at the same time. The satellites also gravitate to their primaries according to the same law that their primaries do to the sun.

Hence, by the law of action and re-action, each body is itself the centre of an attractive force extending indefinitely in space, whence proceed all the mutual disturbances which render the celestial motions so complicated, and their investigation so difficult.

The gravitation of matter, directed to a centre, and attracting directly as the mass and inversely as the square of the distance, does not belong to it when considered in mass only; particle acts on particle according to the same law when at sensible distances from each other. If the sun acted on the centre of the earth without attracting each of its particles, the tides would be very much greater than they now are; and would also, in other respects, be very different. The gravitation of the earth to the sun results from the gravitation of all its particles, which, in their turn, attract the sun in the ratio of their respective masses. There is a reciprocal action likewise between the earth and every particle at its surface; were this not the case, and were any portion of the earth, however small, to attract another portion, and not be itself attracted, the centre of gravity of the earth would be moved in space by this action, which is impossible.

The forms of the planets result from the reciprocal attraction of their component particles. A detached fluid mass, if at rest, would assume the form of a sphere, from the reciprocal attraction of its particles; but if the mass revolves about an axis, it becomes flattened at the poles, and bulges at the equator, in consequence of the centrifugal force arising from the velocity of rotation,—for the centrifugal force diminishes the gravity of the particles at the equator, and equilibrium can only exist where these two forces are balanced by an increase of gravity; therefore, as the attractive force is the same in all particles at equal distances from the centre of a sphere, the equatorial particles would recede from the centre, till their increase in number balanced the centrifugal force by their attraction: consequently, the sphere would become an oblate spheroid; and a fluid partially or entirely covering a solid, as the ocean and atmosphere cover the earth, must assume that form in order to remain in equilibrio. The surface of the sea is therefore spheroidal, and the surface of the earth only deviates from that figure where it rises above, or sinks below, the level of the sea; but the deviation is so small that it is unimportant when compared with the magnitude of the earth—for the mighty chain of the Andes, and the yet more lofty Himalaya, bear about the same proportion to the earth that a grain of sand does to a globe three feet in diameter. Such is the form of the earth and planets; but the compression or flattening at their poles is so small, that even Jupiter, whose rotation is the most rapid, and therefore the most elliptical of the planets, may, from his great distance, be regarded as spherical. Although the planets attract each other as if they were spheres, on account of their distances, yet the satellites are near enough to be sensibly affected in their motions by the forms of their primaries. The moon, for example, is so near the earth, that the reciprocal attraction between each of her particles, and each of the particles in the prominent mass at the terrestrial equator, occasions considerable disturbances in the motions of both bodies: for the action of the moon, on the matter at the earth's equator, produces a nutation in the axis of rotation, and the reaction of that matter on the moon is the cause of a corresponding nutation in the lunar orbit.

If a sphere, at rest in space, receive an impulse passing through its centre of gravity, all its parts will move with an equal velocity in a straight line; but if the impulse does not pass through the centre of gravity, its particles, having unequal velocities, will have a rotatory motion at the same time that it is translated in space. These motions are independent of one another; so that a contrary impulse, passing through its centre of gravity, will impede its progress, without interfering with its rotation.

As the sun rotates about an axis, it seems probable, if an impulse in a contrary direction has not been given to his centre of gravity, that he moves in space, accompanied by all those bodies which compose the solar system—a circumstance which would in no way interfere with their relative motions; for, in consequence of the principle that force is proportional to velocity, the reciprocal attractions of a system remain the same, whether its centre of gravity be at rest, or moving uniformly in space. It is computed that had the earth received its motion from a single impulse, such impulse must have passed through a point about twenty-five miles from its centre.

Since the motions of rotation and translation of the planets are independent of each other, though probably communicated by the same impulse, they form separate subjects of investigation.

SECTION III.

A planet moves in its elliptical orbit with a velocity varying every instant, in consequence of two forces, one tending to the centre of the sun, and the other in the direction of a tangent to its orbit, arising from the primitive impulse given at the time when it was launched into space: should the force in the tangent cease, the planet would fall to the sun by its gravity; were the sun not to attract it, the planet would fly off in the tangent. Thus, when the planet is in aphelion, or at the point where the orbit is farthest from the sun, his action overcomes the planet's velocity, and brings it towards him with such an accelerated motion, that, at last, it overcomes the sun's attraction, and, shooting past him, gradually decreases in velocity, until it arrives at the aphelion where the sun's attraction again prevails. In this motion the *radii vectores*, or imaginary lines joining the centres of the sun and the planets, pass over equal areas in equal times.

If the planets were attracted by the sun only, this would ever be their course; and because his action is proportional to his mass, which is much larger than that of all the planets put together, the elliptical is the nearest approximation to their true motions, which are extremely complicated, in consequence of their mutual attraction, so that they do not move in any known or symmetrical curve, but in paths now approaching to, now receding from, the elliptical form; and their radii vectores do not describe areas exactly proportional to the time. Thus the areas become a test of disturbing forces.

To determine the motion of each body, when disturbed by all the rest, is beyond the power of analysis; it is, therefore, necessary to estimate the disturbing action of one planet at a time, whence the celebrated problem of the three bodies, originally applied to the moon, the earth, and the sun—namely, the masses being given of three bodies projected from three given points, with velocities given both in quantity and direction; and, supposing the bodies to gravitate to one another with forces that are directly as their masses and inversely as the squares of the distances, to find the lines described by these bodies, and their positions at any given instant.

By this problem the motions of translation of all the celestial bodies are determined. It is an extremely difficult one, and would be infinitely more so, if the disturbing action were not very small when compared with the central force. As the disturbing influence of each body may be found separately, it is assumed that the action of the whole system, in disturbing any one planet, is equal to the sum of all the particular disturbances it experiences, on the general mechanical principle, that the sum of any number of small oscillations is nearly equal to their simultaneous and joint effect.

On account of the reciprocal action of matter, the stability of the system depends upon the intensity of the primitive momentum of the planets, and the ratio of their masses to that of the sun—for the nature of the conic sections in which the celestial bodies move, depends upon the velocity with which they were first propelled in space: had that velocity been such as to make the planets move in orbits of unstable equilibrium, their mutual attractions might have changed them into parabolas, or even hyperbolas, so that the earth and planets might, ages ago, have been sweeping far from our sun through the abyss of space: but as the orbits differ very little from circles, the momentum of the planets, when projected, must have been exactly sufficient to ensure the permanency and stability of the system. Besides, the mass of the sun is vastly greater than that of any planet; and as their inequalities bear the same ratio to their elliptical motions as their masses do to that of the sun, their mutual disturbances only increase or diminish the eccentricities of their orbits by very minute quantities; consequently, the magnitude of the sun's mass is the principal cause of the stability of the system. There is not in the physical world a more splendid example of the adaptation of means to the accomplishment of the end, than is exhibited in the nice adjustment of these forces, at once the cause of the variety and of the order of Nature.

The mean distance of a planet from the sun is equal to half the major axis of its orbit: if, therefore, the planet described a circle round the sun at its mean distance, the motion would be uniform, and the periodic time unaltered, because the planet would arrive at the apsides or extremities of the major axis at the same instant, and would have the same velocity, whether it moved in the circular or elliptical orbit, since the curves coincide in these points; but, in every other part, the elliptical motion would either be faster or slower than the circular or mean motion. The difference between the two is called the equation of the centre, which consequently vanishes at the apsides, and is at its maximum ninety degrees distant from these points, or in quadratures, where it measures the eccentricity of the orbit, so that the place of a planet in its elliptical orbit is obtained by adding or subtracting the equation of the centre to or from its mean motion.

The orbits of the planets have a very small inclination to the plane of the ecliptic in which the earth moves; and, on that account, astronomers refer their motions to this plane at a given epoch as a known and fixed position. The paths of the planets, when their mutual disturbances are omitted, are ellipses, nearly approaching to circles, whose planes, slightly inclined to the ecliptic, cut it in straight lines passing through the centre of the sun; the points where an orbit intersects the plane of the ecliptic are its nodes. The ascending node of the lunar orbit, for example, is the point in which the moon rises above the plane of the ecliptic in going towards the north; and her descending node is that in which she sinks below the same plane in moving towards the south. The orbits of the recently discovered planets deviate more from the ecliptic than those of the ancient planets: that of Pallas, for instance, has an inclination of 35° to it; on which account it is more difficult to determine their motions. These little planets have no sensible effect in disturbing the rest, though their own motions are rendered very irregular by the proximity of Jupiter and Saturn.

SECTION IV.

The planets are subject to disturbances of two kinds, both resulting from the constant operation of their reciprocal attraction; one kind, depending upon their positions with regard to each other, begins from zero, increases to a maximum, decreases and becomes zero again, when the planets return to the same relative positions. In consequence of these, the disturbed planet is sometimes drawn away from the sun, sometimes brought nearer to him; at one time it is drawn above the plane of its orbit, at another time below it, according to the position of the disturbing body. All such changes, being accomplished in short periods, some in a few months, others in years, or in hundreds of years, are denominated Periodic Inequalities.

The inequalities of the other kind, though occasioned likewise by the disturbing energy of the planets, are entirely independent of their relative positions: they depend upon the relative positions of the orbits alone, whose forms and places in space are altered by very minute quantities in immense periods of time, and are, therefore, called Secular Inequalities.

In consequence of the latter kind of disturbances, the apsides, or extremities of the major axes of all the orbits, have a direct but variable motion in space, excepting those of the orbit of Venus, which are retrograde; and the lines of the nodes move with a variable velocity in a contrary direction. The motions of both are extremely slow; it requires more than 114755 years for the major axis of the earth's orbit to accomplish a sidereal revolution, that is, to return to the same stars; and 21067 years to complete its tropical motion, or to return to the same equinox. The major axis of Jupiter's orbit requires no less than 200610 years to perform its sidereal revolution, and 22748 years to accomplish its tropical revolution, from the disturbing action of Saturn alone. The periods in which the nodes revolve are also very great. Besides these, the inclination and eccentricity of every orbit are in a state of perpetual but slow change. At the present time the inclinations of all the orbits are decreasing, but so slowly that the inclination of Jupiter's orbit is only about six minutes less now than it was in the age of Ptolemy. The terrestrial eccentricity is decreasing at the rate of about 41·44 miles annually; and, if it were to decrease equably, it would be 37527 years before the earth's orbit became a circle. But, in the midst of all these vicissitudes, the major axes and mean motions of the planets remain permanently independent of secular changes; they are so connected by Kepler's law of the squares of the periodic times being proportional to the cubes of the mean distances of the planets from the sun, that one cannot vary without affecting the other.

With the exception of these two elements, it appears that all the bodies are in motion, and every orbit in a state of perpetual change. Minute as these changes are, they might be supposed to accumulate in the course of ages sufficiently to derange the whole order of nature, to alter the relative positions of the planets, to put an end to the vicissitudes of the seasons, and to bring about collisions which would involve our whole system, now so harmonious, in chaotic confusion. It is natural to inquire what proof exists that nature will be preserved from such a catastrophe? Nothing can be known from observation, since the existence of the human race has occupied comparatively but a point in duration, while these vicissitudes embrace myriads of ages. The proof is simple and convincing. All the variations of the solar system, secular as well as periodic, are expressed analytically by the sines and cosines of circular arcs, which increase with the time; and, as a sine or cosine can never exceed the radius, but must oscillate between zero and unity; however much the time may increase, it follows that when the variations have, by slow changes, accumulated, in however long a time, to a maximum, they decrease, by the same slow degrees, till they arrive at their smallest value, and again begin a new course, thus for ever oscillating about a mean value. This, however, would not be the case if the planets moved in a resisting medium, for then both the eccentricity and the major axes of the orbits would vary with the time, so that the stability of the system would be ultimately destroyed. The existence of such a fluid is now clearly proved; and although it is so extremely rare that hitherto its effects on the motions of the planets have been altogether insensible, there can be no doubt that, in the immensity of time, it will modify the forms of the planetary orbits, and may at last even cause the destruction of our system, which in itself contains no principle of decay.

Three circumstances have generally been supposed necessary to prove the stability of the system: the small eccentricities of the planetary orbits, their small inclinations, and the revolutions of all the bodies, as well planets as satellites, in the same direction. These, however, though sufficient, are not necessary conditions; the periodicity of the terms in which the inequalities are expressed is enough to assure us that, though we do not know the extent of the limits, nor the period of that grand cycle which probably embraces millions of years, yet they never will exceed what is requisite for the stability and harmony of the whole, for the preservation of which every circumstance is so beautifully and wonderfully adapted.

The plane of the ecliptic itself, though assumed to be fixed at a given epoch for the convenience of astronomical computation, is subject to a minute secular variation of 45″·7, occasioned by the reciprocal action of the planets; but, as this is also periodical, and cannot exceed 3°, the terrestrial equator, which is inclined to it at an angle of about 23° 27′ 34″·5, will never coincide with the plane of the ecliptic: so there never can be perpetual spring. The rotation of the earth is uniform;

therefore day and night, summer and winter, will continue their vicissitudes while the system endures, or is undisturbed by foreign causes.

> Yonder starry sphere
> Of planets, and of fixed, in all her wheels
> Resembles nearest mazes intricate,
> Eccentric, intervolved, yet regular,
> Then most, when most irregular they seem.

The stability of our system was established by La Grange: 'a discovery,' says Professor Playfair, 'that must render the name for ever memorable in science, and revered by those who delight in the contemplation of whatever is excellent and sublime.' After Newton's discovery of the mechanical laws of the elliptical orbits of the planets, La Grange's discovery of their periodical inequalities is, without doubt, the noblest truth in physical astronomy; and, in respect of the doctrine of final causes, it may be regarded as the greatest of all.

Notwithstanding the permanency of our system, the secular variations in the planetary orbits would have been extremely embarrassing to astronomers when it became necessary to compare observations separated by long periods. The difficulty was in part obviated, and the principle for accomplishing it established, by La Place; but it has since been extended by M. Poinsot; it appears that there exists an invariable plane passing through the centre of gravity of the system, about which the whole oscillates within very narrow limits, and that this plane will always remain parallel to itself, whatever changes time may induce in the orbits of the planets, in the plane of the ecliptic, or even in the law of gravitation; provided only that our system remains unconnected with any other. The position of the plane is determined by this property —that if each particle in the system be multiplied by the area described upon this plane in a given time, by the projection of its radius vector about the common centre of gravity of the whole, the sum of all these products will be a maximum. La Place found that the plane in question is inclined to the ecliptic at an angle of nearly 1° 35′ 31″, and that, in passing through the sun, and about midway between the orbits of Jupiter and Saturn, it may be regarded as the equator of the solar system, dividing it into two parts, which balance one another in all their motions. This plane of greatest inertia, by no means peculiar to the solar system, but existing in every system of bodies submitted to their mutual attractions only, always maintains a fixed position, whence the oscillations of the system may be estimated through unlimited time. Future astronomers will know, from its immutability or variation, whether the sun and his attendants are connected or not with the other systems of the universe. Should there be no link between them, it may be inferred, from the rotation of the sun, that the centre of gravity of the system situate within his mass describes a straight line in this invariable plane or great equator of the solar system, which, unaffected by the changes of time, will maintain its stability through endless ages. But if the fixed stars, comets, or any unknown and unseen bodies, affect our sun and planets, the nodes of this plane will slowly recede on the plane of that immense orbit which the sun may describe about some most distant centre, in a period which it transcends the powers of man to determine. There is every reason to believe that this is the case; for it is more than probable that, remote as the fixed stars are, they in some degree influence our system, and that even the invariability of this plane is relative, only appearing fixed to creatures incapable of estimating its minute and slow changes during the small extent of time and space granted to the human race. 'The development of such changes,' as M. Poinsot justly observes, 'is similar to an enormous curve, of which we see so small an arc that we imagine it to be a straight line.' If we raise our views to the whole extent of the universe, and consider the stars, together with the sun, to be wandering bodies, revolving about the common centre of creation, we may then recognise in the equatorial plane passing through the centre of gravity of the universe, the only instance of absolute and eternal repose.

All the periodic and secular inequalities deduced from the law of gravitation are so perfectly confirmed by observation, that analysis has become one of the most certain means of discovering the planetary irregularities, either when they are too small, or too long in their periods, to be detected by other methods. Jupiter and Saturn, however, exhibit inequalities which for a long time seemed discordant with that law. All observations, from those of the Chinese and Arabs down to the present day, prove that for ages the mean motions of Jupiter and Saturn have been affected by a great inequality of a very long period, forming an apparent anomaly in the theory of the planets. It was long known by observation that five times the mean motion of Saturn is nearly equal to twice that of Jupiter; a relation which the sagacity of La Place perceived to be the cause of a periodic irregularity in the mean motion of each of these planets, which completes its period in nearly 929 years, the one being retarded while the other is accelerated; but both the magnitude and period of these quantities vary, in consequence of the secular variations in the elements of the orbits. These inequalities are strictly periodical, since they depend upon the configuration of the two planets; and the theory is perfectly confirmed by observation, which shows that, in the course of twenty centuries, Jupiter's mean motion has been accelerated by about 3° 23′, and Saturn's retarded by 5° 13′.

It might be imagined that the reciprocal action of such planets as have satellites would be different from the influence of those that have none; but the distances of the satellites from their primaries are incomparably less than the distances of the planets from the sun, and from one another; so that the system of a planet and its satellites moves nearly as if all these bodies were united in their common centre of gravity: the action of the sun, however, in some degree disturbs the motion of the satellites about their primary.

SECTION V.

The changes which take place in the planetary system are exhibited on a smaller scale by Jupiter and his satellites: and, as the period requisite for the development of the inequalities of these little moons only extends to a few centuries, it may be regarded as an epitome of that grand cycle which will not be accomplished by the planets in myriads of ages. The revolutions of the satellites about Jupiter are precisely similar to those of the planets about the sun: it is true they are disturbed by the sun, but his distance is so great, that their motions are nearly the same as if they were not under his influence. The satellites, like the planets, were probably projected in elliptical orbits, but the compression of Jupiter's spheroid is very great in consequence of his rapid rotation; and as the masses of the satellites are nearly 100000 times less than that of Jupiter, the immense quantity of prominent matter at his equator must soon have given the circular form observed in the orbits of the first and second satellites, which its superior attraction will always maintain. The third and fourth satellites being farther removed from its influence, move in orbits with a very small eccentricity. The same cause occasions the orbits of the satellites to remain nearly in the plane of Jupiter's equator, on account of which they are always seen nearly in the same line; and the powerful action of that quantity of prominent matter is the reason why the motions of the nodes of these small bodies is so much more rapid than those of the planet. The nodes of the fourth satellite accomplish a tropical revolution in 531 years, while those of Jupiter's orbit require no less than 36261 years,—a proof of the reciprocal attraction between each particle of Jupiter's equator and of the satellites. Although the two first satellites sensibly move in circles, they acquire a small ellipticity from the disturbances they experience.

The orbits of the satellites do not retain a permanent inclination either to the plane of Jupiter's equator or to that of his orbit, but to certain planes passing between the two, and through their intersection; these have a greater inclination to his equator the farther the satellite is removed, owing to the influence of Jupiter's compression, and they have a slow motion corresponding to secular variations in the planes of Jupiter's orbit and equator.

The satellites are not only subject to periodic and secular inequalities from their mutual attraction, similar to those which affect the motions and orbits of the planets, but also to others peculiar to themselves. Of the periodic inequalities arising from their mutual attraction the most remarkable take place in the angular motions of the three nearest to Jupiter, the second of which receives from the first a perturbation similar to that which it produces in the third; and it experiences from the third a perturbation similar to that which it communicates to the first. In the eclipses these two inequalities are combined into one, whose period is 437·659$^{\text{days.}}$ The variations peculiar to the satellites arise from the secular inequalities occasioned by the action of the planets in the form and position of Jupiter's orbit, and from the displacement of his equator. It is obvious that whatever alters the relative positions of the sun, Jupiter, and his satellites, must occasion a change in the directions and intensities of the forces, which will affect the motions and orbits of the satellites. For this reason the secular variations in the eccentricity of Jupiter's orbit, occasion secular inequalities in the mean motions of the satellites, and in the motions of the nodes and apsides of their orbits. The displacement of the orbit of Jupiter, and the variation in the position of his equator, also affect these small bodies. The plane of Jupiter's equator is inclined to the plane of his orbit, so that the action of the sun and of the satellites themselves produces a nutation and precession in his equator, precisely similar to that which takes place in the rotation of the earth, from the action of the sun and moon, whence the protuberant matter at Jupiter's equator is continually changing its position with regard to the satellites, and produces corresponding mutations in their motions; and, as the cause must be proportional to the effect, these inequalities afford the means, not only of ascertaining the compression of Jupiter's spheroid, but they prove that his mass is not homogeneous. Although the apparent diameters of the satellites are too small to be measured, yet their perturbations give the values of their masses with considerable accuracy,—a striking proof of the power of analysis.

A singular law obtains among the mean motions and mean longitudes of the three first satellites. It appears from observation that the mean motion of the first satellite, plus twice that of the third, is equal to three times that of the second; and that the mean longitude of the first satellite, minus three times that of the second, plus twice that of the third, is always equal to two right angles. It is proved by theory, that if these relations had only been approximate when the satellites were first launched into space, their mutual attractions would have established and maintained them, notwithstanding the secular inequalities to which they are liable. They extend to the synodic motions of the satellites, consequently they affect their eclipses, and have a very great influence on their whole theory. The satellites move so nearly in the plane of Jupiter's equator, which has a very small inclination to his orbit, that they are frequently eclipsed by the shadow of the planet. The eclipses take place close to the disc of Jupiter when he is near opposition; but at times the shadow is so projected with regard to the earth, that the third and fourth satellites vanish and reappear on the same side of the disc. These eclipses are in all respects similar to those of the moon; but, occasionally, the satellites eclipse Jupiter, passing like black spots across his surface, and resemble annular eclipses of the sun. The instant of the beginning or end of an eclipse of a satellite marks the same instant of absolute time to all the inhabitants of the earth; therefore, the time of these eclipses observed by a traveller, when compared with the time of the eclipse computed for Greenwich, or any other fixed meridian, gives the difference of the meridians in time, and consequently the longitude of the place of observation. It has required all the refinements of modern instruments to render the eclipses of these remote moons available to the mariner; now, however, that system of bodies invisible to the naked eye, known to man by the aid of science alone, enables him to traverse the ocean, spreading the light of knowledge and the blessings of civilization over the most remote regions, and to return loaded with the productions of another hemisphere. Nor is this all: the eclipses of Jupiter's satellites have been the means of a discovery which, though not so immediately applicable to the wants of man, unfolds one of the properties of light,—that medium without whose cheering influence all the beauties of the creation would have been to us a blank. It is observed, that those eclipses of the first satellite, which happen when Jupiter is near conjunction, are later by $16^m\ 26^s$ than those which take place when the planet is in opposition. But, as Jupiter is nearer to us when in opposition by the whole breadth of the earth's orbit than when in conjunction, this circumstance was attributed to the time employed by the rays of light in crossing the earth's orbit, a distance of about 190 millions of miles; whence it is estimated that light travels at the rate of 190000 miles in one second. Such is its velocity, that the earth, moving at the rate of 19 miles in a second, would take two months to pass through a distance which a ray of light would dart over in eight minutes. The subsequent discovery of the aberration of light confirmed this astonishing result.

Objects appear to be situate in the direction of the rays which proceed from them. Were light propagated instantaneously, every object, whether at rest or in motion, would appear in the direction of these rays; but as light takes some time to travel, we see Jupiter in conjunction, by means of rays that left him $16^m\ 26^s$ before; but, during that time, we have changed our position, in consequence of the motion of the earth in its orbit; consequently we refer Jupiter to a place in which he is not. His true position is in the diagonal of the parallelogram, whose sides are in the ratio of the velocity of light to the velocity of the earth in its orbit, which is as 190000 to 19. In consequence of the aberration of light, the heavenly bodies seem to be in places in which they are not. In fact, if the earth were at rest, rays from a star would pass along the axis of a telescope directed to it: but if the earth were to begin to move in its orbit, with its usual velocity, these rays would strike against the side of the tube; it would, therefore, be necessary to incline the telescope a little, in order to see the star. The angle contained between the axis of the telescope and a line drawn to the true place of the star, is its aberration, which varies in quantity and direction in different parts of the earth's orbit; but as it is only $20''\cdot37$, or $20''\cdot5$, it is insensible in ordinary cases.

The velocity of light deduced from the observed aberration of the fixed stars, perfectly corresponds with that given by the eclipses of the first satellite. The same result, obtained from sources so different, leaves not a doubt of its truth. Many such beautiful coincidences, derived from circumstances apparently the most unpromising and dissimilar, occur in the rest of physical astronomy, and prove dependences which we might otherwise be unable to trace. The identity of the velocity of light, at the distance of Jupiter, and on the earth's surface, shows that its velocity is uniform; and if light consists in the vibrations of an elastic fluid or ether filling space, an hypothesis which accords best with observed phenomena, the uniformity of its velocity shows that the density of the fluid throughout the whole extent of the solar system must be proportional to its elasticity. Among the fortunate conjectures which have been confirmed by subsequent experience, that of Bacon is not the least remarkable. 'It produces in me,' says the restorer of true philosophy, 'a doubt whether the face of the serene and starry heavens be seen at the instant it really exists, or not till some time later; and whether there be not, with respect to the heavenly bodies, a true time and an apparent time, no less than a true place and an apparent place, as astronomers say, on account of parallax. For it seems incredible that the species or rays of the celestial bodies can pass through the immense interval between them and us in an instant, or that they do not even require some considerable portion of time.'

As great discoveries generally lead to a variety of conclusions, the aberration of light affords a direct proof of the motion of the earth in its orbit;

and its rotation is proved by the theory of falling bodies, since the centrifugal force it induces retards the oscillations of the pendulum in going from the pole to the equator. Thus a high degree of scientific knowledge has been requisite to dispel the errors of the senses.

The little that is known of the theories of the satellites of Saturn and Uranus is, in all respects, similar to that of Jupiter. The great compression of Saturn occasions its satellites to move nearly in the plane of its equator. Of the situation of the equator of Uranus we know nothing, nor of his compression; but the orbits of his satellites are nearly perpendicular to the plane of the ecliptic, and by analogy they ought to be in the plane of his equator.

SECTION VI.

Our constant companion, the moon, next claims our attention. Several circumstances concur to render her motions the most interesting, and at the same time the most difficult to investigate of all the bodies of our system. In the solar system planet troubles planet, but in the lunar theory the sun is the great disturbing cause; his vast distance being compensated by his enormous magnitude, so that the motions of the moon are more irregular than those of the planets; and, on account of the great ellipticity of her orbit, and the size of the sun, the approximations to her motions are tedious and difficult beyond what those unaccustomed to such investigations could imagine. Among the innumerable periodic inequalities to which the moon's motion in longitude is liable, the most remarkable are the Evection, the Variation, and the Annual Equation. The forces producing the evection diminish the excentricity of the lunar orbit in conjunction and opposition, and augment it in quadrature. The period of this inequality is less than thirty-two days. Were the increase and diminution always the same, the evection would only depend upon the distance of the moon from the sun; but its absolute value also varies with her distance from the perigee of her orbit. Ancient astronomers, who observed the moon solely with a view to the prediction of eclipses, which can only happen in conjunction and opposition, where the excentricity is diminished by the evection, assigned too small a value to the ellipticity of her orbit. The variation, which is at its maximum when the moon is 45° distant from the sun, vanishes when that distance amounts to a quadrant, and also when the moon is in conjunction and opposition; consequently, that inequality never could have been discovered from the eclipses: its period is half a lunar month. The annual equation arises from

D 2

the moon's motion being accelerated when that of the earth is retarded, and *vice versâ*—for, when the earth is in its perihelion, the lunar orbit is enlarged by the action of the sun; therefore, the moon requires more time to perform her revolution. But, as the earth approaches its aphelion, the moon's orbit contracts, and less time is necessary to accomplish her motion,—its period, consequently, depends upon the time of the year. In the eclipses the annual equation combines with the equation of the centre of the terrestrial orbit, so that ancient astronomers imagined the earth's orbit to have a greater excentricity than modern astronomers assign to it.

The planets disturb the motion of the moon both directly and indirectly; because their action on the earth alters its relative position with regard to the sun and moon, and occasions inequalities in the moon's motion, which are more considerable than those arising from their direct action: for the same reason the moon, by disturbing the earth, indirectly disturbs her own motion. Neither the excentricity of the lunar orbit, nor its mean inclination to the plane of the ecliptic, have experienced any changes from secular inequalities; for, although the mean action of the sun on the moon depends upon the inclination of the lunar orbit to the ecliptic, and that the position of the ecliptic is subject to a secular inequality, yet analysis shows that it does not occasion a secular variation in the inclination of the lunar orbit, because the action of the sun constantly brings the moon's orbit to the same inclination on the ecliptic. The mean motion, the nodes, and the perigee, however, are subject to very remarkable variations.

From an eclipse observed by the Chaldeans at Babylon, on the 19th of March, seven hundred and twenty-one years before the Christian era, the place of the moon is known from that of the sun at the instant of opposition, whence her mean longitude may be found; but the comparison of this mean longitude with another mean longitude, computed back for the instant of the eclipse from modern observations, shows that the moon performs her revolution round the earth more rapidly and in a shorter time now, than she did formerly; and that the acceleration in her mean motion has been increasing from age to age as the square of the time: all ancient and intermediate eclipses confirm this result. As the mean motions of the planets have no secular inequalities, this seemed to be an unaccountable anomaly. It was at one time attributed to the resistance of an etherial medium pervading space, and at another to the successive transmission of the gravitating force; but as La Place proved that neither of these causes, even if they exist, have any influence on the motions of the lunar perigee or nodes, they could not affect the mean motion; a variation in the mean motion from such causes being inseparably connected with variations in the motions of the perigee and nodes. That great mathematician, in studying the theory of Jupiter's satellites, perceived that the secular variation in the elements of Jupiter's orbit, from the action of the planets, occasions corresponding changes in the motions of the satellites, which led him to suspect that the acceleration in the mean motion of the moon might be connected with the secular variation in the excentricity of the terrestrial orbit; and analysis has proved that he assigned the true cause of the acceleration.

If the excentricity of the earth's orbit were invariable, the moon would be exposed to a variable disturbance from the action of the sun, in consequence of the earth's annual revolution; it would however be periodic, since it would be the same as often as the sun, the earth, and the moon returned to the same relative positions: but on account of the slow and incessant diminution in the excentricity of the terrestrial orbit, the revolution of our planet is performed at different distances from the sun every year. The position of the moon with regard to the sun undergoes a corresponding change; so that the mean action of the sun on the moon varies from one century to another, and occasions the secular increase in the moon's velocity called the Acceleration, a name peculiarly appropriate in the present age, and which will continue to be so for a vast number of ages to come; because, as long as the earth's excentricity diminishes, the moon's mean motion will be accelerated, but when the excentricity has passed its minimum, and begins to increase, the mean motion will be retarded from age to age. At present the secular acceleration is about 11″·209, but its effect on the moon's place increases as the square of the time. It is remarkable that the action of the planets thus reflected by the sun to the moon is much more sensible than their direct action, either on the earth or moon. The secular diminution in the excentricity, which has not altered the equation of the centre of the sun by eight minutes since the earliest recorded eclipses, has produced a variation of about 1° 48′ in the moon's longitude, and of 7° 12′ in her mean anomaly.

The action of the sun occasions a rapid but variable motion in the nodes and perigee of the lunar orbit. Though the nodes recede during the greater part of the moon's revolution, and advance during the smaller, they perform their sidereal revolution in 6793·37953 days; and the perigee accomplishes a revolution in 3232·56731 days, or a little more than nine years, notwithstanding its motion is sometimes retrograde and sometimes direct; but such is the difference between the disturbing energy of the sun and that of all the planets put together, that it requires no less than 114755 years for the greater axis of the terrestrial orbit to do the same. It is evident that the same secular variation which changes the sun's distance from the earth, and occasions the acceleration in the moon's mean motion, must affect the nodes and perigee; and it consequently appears, from theory as well as observation, that both these elements are subject to a secular inequality arising from the variation in the excentricity of the earth's orbit, which connects them with the Acceleration, so that both are retarded when the mean motion is anticipated. The secular variations in these three elements are in the ratio of the numbers 3, 0·735, and 1; whence the three motions of the moon, with regard to the sun, to her perigee, and to her nodes, are continually accelerated, and their secular equations are as the numbers 1, 4, and 0·265, or, according to the most recent investigations, as 1, 4·6776, and 0·391. A comparison of ancient eclipses observed by the Arabs, Greeks, and Chaldeans, imperfect as they are, with modern observations, perfectly confirms these results of analysis. Future ages will develop these great inequalities, which at some most distant period will amount to many circumferences. They are indeed periodic; but who shall tell their period? Millions of years must elapse before that great cycle is accomplished; but 'such changes, though rare in time, are frequent in eternity.'

The moon is so near, that the excess of matter at the earth's equator occasions periodic variations in her longitude, and also that remarkable inequality in her latitude already mentioned as a nutation in the lunar orbit, which diminishes its inclination to the ecliptic when the moon's ascending node coincides with the equinox of spring, and augments it when that node coincides with the equinox of autumn. As the cause must be proportional to the effect, a comparison of these inequalities, computed from theory, with the same given by observation, shows that the compression of the terrestrial spheroid, or the ratio of the difference between the polar and equatorial diameters, to the diameter of the equator, is $\frac{1}{305 \cdot 05}$. It is proved analytically that, if a fluid mass of homogeneous matter, whose particles attract each other inversely as the square of the distance, were to revolve about an axis as the earth does, it would assume the form of a spheroid whose compression is $\frac{1}{230}$, whence it appears that the earth is not homogeneous, but decreases in density from its centre to its circumference. Thus the moon's eclipses show the earth to be round, and her inequalities not only determine the form, but the internal structure of our planet; results of analysis which could not have been anticipated. Similar inequalities in the motions of Jupiter's satellites prove that his mass is not homogeneous, and that his compression is $\frac{1}{13 \cdot 8}$. His equatorial diameter exceeds his polar diameter by about 6230 miles.

The phases of the moon, which vary from a slender silvery crescent soon after conjunction to a complete circle of light in opposition, decrease by the same degrees till the moon is again enveloped in the morning beams of the sun. These changes regulate the returns of the eclipses; those of the sun can only happen in conjunction, when the moon, coming between the earth and the sun, intercepts his light; and those of the moon are occasioned by the earth intervening between the sun and moon when in opposition. As the earth is opaque and nearly spherical, it throws a conical shadow on the side of the moon opposite to the sun, the axis of which passes through the centres of the sun and earth. The length of the shadow terminates at the point where the apparent diameters of the sun and earth would be the same. When the moon is in opposition, and at her mean distance, the diameter of the sun would be seen from her centre under an angle of 1918″·1; and that of the earth would appear under an angle of 6908″·3; so that the length of the shadow is at least three times and a half greater than the distance of the moon from the earth, and the breadth of the shadow, where it is traversed by the moon, is about eight-thirds of the lunar diameter. Hence the moon would be eclipsed every opposition, were it not for the inclination of her orbit to the plane of the ecliptic, in consequence of which the moon in opposition is either above or below the cone of the earth's shadow, except when in or near her nodes; her position with regard to them occasions all the varieties in the lunar eclipses. Every point of the moon's surface successively loses the light of different parts of the sun's disc before being eclipsed. Her brightness therefore gradually diminishes before she plunges into the earth's shadow. The breadth of the space occupied by the penumbra is equal to the apparent diameter of the sun, as seen from the centre of the moon. The mean duration of a revolution of the sun, with regard to the node of the lunar orbit, is to the duration of a synodic revolution of the moon as 223 to 19; so that, after a period of 223 lunar months, the sun and moon would return to the same relative position to the node of the moon's orbit, and therefore the eclipses would recur in the same order, were not the periods altered by irregularities in the motions of the sun and moon. In lunar eclipses, our atmosphere refracts the sun's rays which pass through it, and bends them all round into the cone of the earth's shadow; and as the horizontal refraction surpasses half the sum of the solar and lunar parallaxes, that is, half the sum of the semidiameters of the sun and moon, divided by their mutual distance, the centre of the lunar disc, supposed to be in the axis of the shadow, would receive the rays from the same point of the sun, round all sides of the earth, so that it would be more illuminated than in full moon, if the greater portion of the light were not absorbed by the atmosphere. Instances are recorded where this feeble light has been entirely absorbed, so that the moon has altogether disappeared in her eclipses.

The sun is eclipsed when the moon intercepts his rays. The moon, though incomparably smaller than the sun, is so much nearer the earth, that her apparent diameter differs but little from his, but both are liable to such variations, that they alternately surpass one another. Were the eye of a spectator in the same straight line with the centres of the sun and moon, he would see the sun eclipsed. If the apparent diameter of the moon surpassed that of the sun, the eclipse would be total; if it were less, the observer would see a ring of light round the disc of the moon, and the eclipse would be annular. If the centre of the moon should not be in the straight line joining the centres of the sun and the eye of the observer, the moon might only eclipse a part of the sun. The variation, therefore, in the distances of the sun and moon from the centre of the earth, and of the moon from her node at the instant of conjunction, occasions great varieties in the solar eclipses. Besides, the height of the moon above the horizon changes her apparent diameter, and may augment or diminish the apparent distances of the centres of the sun and moon, so that an eclipse of the sun may occur to the inhabitants of one country, and not to those of another. In this respect the solar eclipses differ from the lunar, which are the same for every part of the earth where the sun and moon are above the horizon. In solar eclipses, the light reflected by the atmosphere diminishes the obscurity they produce; even in total eclipses the higher part of the atmosphere is enlightened by a part of the sun's disc, and reflects its rays to the earth. The whole disc of the new moon is frequently visible from atmospheric reflection.

Planets sometimes eclipse one another. On the 17th of May, 1737, Mercury was eclipsed by Venus near their inferior conjunction: Mars passed over Jupiter on the 9th of January, 1591, and on the 13th of October, 1825, the moon eclipsed Saturn. These phenomena, however, happen very seldom, because all the planets, or even a part of them, are very rarely seen in conjunction at once; that is, in the same part of the heavens at the same time. More than 2500 years before our era, the five great planets were in conjunction. On the 15th of September, 1186, a similar assemblage took place between the constellations of Virgo and Libra; and in 1801, the Moon, Jupiter, Saturn, and Venus were united in the heart of the Lion. These conjunctions are so rare, that Lalande has computed that more than seventeen millions of millions of years separate the epochs of the contemporaneous conjunctions of the six great planets.

The motions of the moon have now become of more importance to the navigator and geographer than those of any other heavenly body, from the precision with which the longitude is determined by the occultations of stars and lunar distances. The occultation of a star by the moon is a phenomenon of frequent occurrence: the moon seems to pass over the star, which almost instantaneously vanishes at one side of her disc, and after a short time as suddenly reappears on the other; and a lunar distance is the observed distance of the moon from the sun, or from a particular star or planet, at any instant. The lunar theory is brought to such perfection, that the times of these phenomena, observed under any meridian, when compared with those computed for Greenwich in the Nautical Almanac, give the longitude of the observer within a few miles. The accuracy of that work is obviously of extreme importance to a maritime nation: we have reason to hope that the new Ephemeris, now in preparation, will be by far the most perfect work of the kind that ever has been published.

From the lunar theory, the mean distance of the sun from the earth, and thence the whole dimensions of the solar system, are known; for the forces which retain the earth and moon in their orbits are respectively proportional to the radii vectores of the earth and moon, each being divided by the square of its periodic time; and as the lunar theory gives the ratio of the forces, the ratio of the distances of the sun and moon from the earth is obtained; whence it appears that the sun's mean distance from the earth is nearly 396 times greater than that of the moon. The method, however, of finding the absolute distances of the celestial bodies in miles, is in fact the same with that employed in measuring the distances of terrestrial objects. From the extremities of a known base, the angles which the visual rays from the object form with it are measured; their sum subtracted from two right angles gives the angle opposite the base; therefore, by trigonometry, all the angles and sides of the triangle may be computed—consequently the distance of the object is found. The angle under which the base of the triangle is seen from the object is the parallax of that object; it evidently increases and decreases with the distance; therefore the base must be very great indeed to be visible at all from the celestial bodies. The globe itself, whose dimensions are obtained by actual admeasurement, furnishes a standard of measures, with which we compare the distances, masses, densities, and volumes of the sun and planets.

SECTION VII.

THE theoretical investigation of the figure of the earth and planets is so complicated, that neither the geometry of Newton nor the refined analyses of La Place have attained more than an approximation: it is only within a few years that a complete and finite solution of that difficult problem has been accomplished by our distinguished countryman Mr. Ivory. The investigation has been conducted by successive steps, beginning with a simple case, and then proceeding to the more difficult; but in all, the forces which occasion the revolutions of the earth and planets are omitted, because, by acting equally upon all the particles, they do not disturb their mutual relations. A fluid mass of uniform density, whose particles mutually gravitate to each other, will assume the form of a sphere when at rest; but if the sphere begins to revolve, every particle will describe a circle, having its centre in the axis of revolution; the planes of all these circles will be parallel to one another, and perpendicular to the axis, and the particles will have a tendency to fly from that axis in consequence of the centrifugal force arising from the velocity of rotation. The force of gravity is everywhere perpendicular to the surface, and tends to the interior of the fluid mass, whereas the centrifugal force acts perpendicularly to the axis of rotation, and is directed to the exterior; and as its intensity diminishes with the distance from the axis of rotation, it decreases from the equator to the poles, where it ceases. Now it is clear that these two forces are in direct opposition to each other in the equator alone, and that gravity is there diminished by the whole effect of the centrifugal force, whereas, in every other part of the fluid, the centrifugal force is resolved into two parts, one of which, being perpendicular to the surface, diminishes the force of gravity; but the other, being at a tangent to the surface, urges the particles towards the equator, where they accumulate till their numbers compensate the diminution of gravity, which makes the mass bulge at the equator and become flattened at the poles. It appears, then, that the influence of the centrifugal force is most powerful at the equator, not only because it is actually greater there than elsewhere, but because its whole effect is employed in diminishing gravity, whereas, in every other point of the fluid mass, it is only a resolved part that is so employed. For both these reasons it gradually decreases towards the poles, where it ceases. On the contrary, gravity is least at the equator, because the particles are farther from the centre of the mass, and increases towards the poles, where it is greatest. It is evident, therefore, that, as the centrifugal force is much less than the force of gravity,—gravitation, which is the difference between the two, is least at the equator, and continually increases towards the poles, where it is a maximum. On these principles Sir Isaac Newton proved that a homogeneous fluid mass in rotation assumes the form of an ellipsoid of revolution, whose compression is $\frac{1}{230}$. Such, however, cannot be the form of the earth, because the strata increase in density towards the centre. The lunar inequalities also prove the earth to be so constructed; it was requisite, therefore, to consider the fluid mass to be of variable density. Including this condition, it has been found that the mass, when in rotation, would still assume the form of an ellipsoid of revolution; that the particles of equal density would arrange themselves in concentric elliptical strata, the most dense being in the centre; but that the compression would be less than in the case of the homogeneous fluid. The compression is still less when the mass is considered to be, as it actually is, a solid nucleus, decreasing regularly in density from the centre to the surface, and partially covered by the ocean, because the solid parts, by their cohesion, nearly destroy that part of the centrifugal force which gives the particles a tendency to accumulate at the equator, though not altogether; otherwise the sea, by the superior mobility of its particles, would flow towards the equator and leave the poles dry: besides, it is well known that the continents at the equator are more elevated than they are in higher latitudes. It is also necessary for the equilibrium of the ocean, that its density should be less than the mean density of the earth, otherwise the continents would be perpetually liable to inundations from storms and other causes. On the whole, it appears from theory that a horizontal line passing round the earth, through both poles, must be nearly an ellipse, having its major axis in the plane of the equator, and its minor axis coinciding with the axis of the earth's rotation. The intensity of the centrifugal force is measured by the deflection of any point from the tangent in a second, and is determined from the known velocity of the earth's rotation: the force of gravitation at any place is measured by the descent of a heavy body in the first second of its fall. At the equator the centrifugal force is equal to the 289th part of gravity, and diminishes towards the poles as the cosine of the latitude, for the angle between the directions of these two forces, at any point of the surface, is equal to its latitude. But whatever the constitution of the earth and planets may be, analysis proves that, if the intensity of gravitation at the equator be taken equal to unity, the sum of the compression of the ellipsoid and the whole increase of gravitation, from the equator to the pole, is equal to five-halves of the ratio of the centrifugal force to gravitation at the equator. This quantity, with regard to the earth, is $\frac{5}{2}$ of $\frac{1}{289}$, or $\frac{1}{115 \cdot 2}$; consequently the compression of the earth is equal to $\frac{1}{115 \cdot 2}$, diminished by the whole increase of gravitation, so that its form will be known, if the whole increase of gravitation, from the equator to the pole, can be determined by experiment. But there is another method of ascertaining the figure of our planet. It is easy to show, in a spheroid whose strata are elliptical, that the increase in the length of the radii, the decrease of gravitation, and the increase in the lengths of the arcs of the meridian, corresponding to angles of one degree, from the pole to the equator, are proportional to the square of the cosine of the latitude. These quantities are so connected with the ellipticity of the spheroid, that the total increase in the length of the radii is equal to the compression, and the total diminution in the length of the arcs is equal to the compression multiplied by three times the length of an arc of one degree at the equator. Hence, by measuring the meridian curvature of the earth, the compression, and consequently its figure, become known. This, indeed, is assuming the earth to be an ellipsoid of revolution, but the actual measurement of the globe will show how far it corresponds with that solid in figure and constitution.

The courses of the great rivers, which are in general navigable to a considerable extent, prove that the curvature of the land differs but little from that of the ocean; and as the heights of the mountains and continents are inconsiderable when compared with the magnitude of the earth, its figure is understood to be determined by a surface at every point perpendicular to the direction of gravitation, or of the plumb-line, and is the same which the sea would have if it were continued all round the earth beneath the continents. Such is the figure that has been measured in the following manner:—

A terrestrial meridian is a line passing through both poles, all the points of which have their noon contemporaneously. Were the lengths and curvatures of different meridians known, the figure of the earth might be determined; but the length of one degree is sufficient to give the figure of the earth, if it be measured on different meridians, and in a variety of latitudes; for if the earth were a sphere, all degrees would be of the same length, but if not, the lengths of the degrees will be greatest where the curvature is least, and will be greater exactly in proportion as the curvature is less; a comparison of the lengths of the degree in different parts of the earth's surface will therefore determine its size and form.

An arc of the meridian may be measured by observing the latitude of its extreme points, and then measuring the distance between them in feet or fathoms: the distance thus determined on the surface of the earth, divided by the degrees and parts of a degree contained in the difference of the latitudes, will give the exact length of one degree, the difference of the latitudes being the angle contained between the verticals at the extremities of the arc. This would be easily accomplished were the distance unobstructed, and on a level with the sea; but on account of the innumerable obstacles on the surface of the earth, it is necessary to connect the extreme points of the arc by a series of triangles, the sides and angles of which are either measured or computed, so that the length of the arc is ascertained with much laborious computation. In consequence of the irregularities of the surface, each triangle is in a different plane; they must therefore be reduced by computation to what they would have been, had they been measured on the surface of the sea; and as the earth may in this case be esteemed spherical, they require a correction to reduce them to spherical triangles.

Arcs of the meridian have been measured in a variety of latitudes north and south, as well as arcs perpendicular to the meridian. From these measurements it appears that the lengths of the degrees increase from the equator to the poles, nearly in proportion to the square of the sine of the latitude; consequently the convexity of the earth diminishes from the equator to the poles.

Were the earth an ellipsoid of revolution, the

meridians would be ellipses whose lesser axes would coincide with the axis of rotation, and all the degrees measured between the pole and the equator would give the same compression when combined two and two. That, however, is far from being the case. Scarcely any of the measurements give exactly the same results, chiefly on account of local attractions, which cause the plumb-line to deviate from the vertical. The vicinity of mountains has that effect; but one of the most remarkable, though not unprecedented, anomalies takes place in the plains in the north of Italy, where the action of some dense subterraneous matter causes the plumb-line to deviate seven or eight times more than it did from the attraction of Chimborazo during the experiments of Bouguer, while measuring a degree of the meridian at the equator. In consequence of this local attraction, the degrees of the meridian in that part of Italy seem to increase towards the equator through a small space, instead of decreasing, as if the earth was drawn out at the poles, instead of being flattened.

Many other discrepancies occur, but from the mean of the five principal measurements of arcs in Peru, India, France, England, and Lapland, Mr. Ivory has deduced that the figure which most nearly follows this law is an ellipsoid of revolution whose equatorial radius is 3962·824 miles, and the polar radius 3949·585 miles; the difference, or 13·239 miles, divided by the equatorial radius, is $\frac{1}{298\cdot33}$ from arcs of the meridian, $\frac{1}{282\cdot90}$ from the pendulum, and the true compression is $\frac{1}{290\cdot615}$: this fraction is called the compression of the earth, because, according as it is greater or less, the terrestrial ellipsoid is more or less flattened at the poles; it does not differ much from that given by the lunar inequalities. If we assume the earth to be a sphere, the length of a degree of the meridian is $69\frac{1}{22}$ British miles; therefore 360 degrees, or the whole circumference of the globe, is 24856 miles, and the diameter, which is something less than a third of the circumference, is about 7912 or 8000 miles nearly. Eratosthenes, who died 194 years before the Christian era, was the first to give an approximate value of the earth's circumference, by the measurement of an arc between Alexandria and Syene.

The other method of finding the figure of the earth is totally independent of either of the preceding. If the earth were a homogeneous sphere without rotation, its attraction on bodies at its surface would be everywhere the same; if it be elliptical and of variable density, the force of gravity, theoretically, ought to increase from the equator to the pole, as unity *plus* a constant quantity multiplied into the square of the sine of the latitude; but for a spheroid in rotation, the centrifugal force varies, by the laws of mechanics, as the square of the sine of the latitude, from the equator, where it is greatest, to the pole, where it vanishes; and as it tends to make bodies fly off the surface, it diminishes the force of gravity by a small quantity. Hence, by gravitation, which is the difference of these two forces, the fall of bodies ought to be accelerated from the equator to the poles, proportionably to the square of the sine of the latitude; and the weight of the same body ought to increase in that ratio. This is directly proved by the oscillations of the pendulum; for if the fall of bodies be accelerated, the oscillations will be more rapid; and in order that they may always be performed in the same time, the length of the pendulum must be altered. By numerous and careful experiments, it is proved that a pendulum which oscillates 86400 times in a mean day at the equator will do the same at every point of the earth's surface, if its length be increased progressively to the pole, as the square of the sine of the latitude.

From the mean of these it appears that the whole decrease of gravitation from the poles to the equator is 0·001457, which, subtracted from $\frac{1}{115{\cdot}2}$, shows that the compression of the terrestrial spheroid is about $\frac{1}{282{\cdot}90}$, which does not differ much from that given by the lunar inequalities, and from the arcs in the direction of the meridian, as well as those perpendicular to it. The near coincidence of these three values, deduced by methods so entirely independent of each other, shows that the mutual tendencies of the centres of the celestial bodies to one another, and the attraction of the earth for bodies at its surface, result from the reciprocal attraction of all their particles. Another proof may be added: the nutation of the earth's axis, and the precession of the equinoxes, are occasioned by the action of the sun and moon on the protuberant matter at the earth's equator; and although these inequalities do not give the absolute value of the terrestrial compression, they show that the fraction expressing it is comprised between the limits $\frac{1}{279}$ and $\frac{1}{578}$.

It might be expected that the same compression should result from each, if the different methods of observation could be made without error. This, however, is not the case; for, after allowance has been made for every cause of error, such discrepances are found, both in the degrees of the meridian and in the length of the pendulum, as show that the figure of the earth is very complicated; but they are so small, when compared with the general results, that they may be disregarded. The compression deduced from the mean of the whole appears to be about $\frac{1}{290 \cdot 615}$; that given by the lunar theory has the advantage of being independent of the irregularities of the earth's surface and of local attractions. The regularity with which the observed variation in the length of the pendulum follows the law of the square of the sine of the latitude proves the strata to be elliptical and symmetrically disposed round the centre of gravity of the earth, which affords a strong presumption in favour of its original fluidity. It is remarkable how little influence the sea has on the variation of the lengths of the arcs of the meridian or on gravitation, neither does it much affect the lunar inequalities, from its density being only about a fifth of the mean density of the earth. For, if the earth were to become fluid after being stripped of the ocean, it would assume the form of an ellipsoid of revolution whose compression is $\frac{1}{304 \cdot 8}$, which differs very little from that determined by observation, and proves, not only that the density of the ocean is inconsiderable, but that its mean depth is very small. There may be profound cavities in the bottom of the sea, but its mean depth probably does not much exceed the mean height of the continents and islands above its level. On this account, immense tracts of land may be deserted or overwhelmed by the ocean, as appears really to have been the case, without any great change in the form of the terrestrial spheroid. The variation in the length of the pendulum was first remarked by Richter in 1672, while observing transits of the fixed stars across the meridian at Cayenne, about five degrees north of the equator. He found that his clock lost at the rate of $2^{m.}$ $28^{s.}$ daily, which induced him to determine the length of a pendulum beating seconds in that latitude; and repeating the experiments on his return to Europe, he found the seconds pendulum at Paris to be more than the twelfth of an inch longer than that at Cayenne. The form and size of the earth being determined, it furnishes a standard of measure with which the dimensions of the solar system may be compared.

SECTION VIII.

THE parallax of a celestial body is the angle under which the radius of the earth would be seen if viewed from the centre of that body; it affords the means of ascertaining the distances of the sun, moon, and planets. Suppose, when the moon is in the horizon at the instant of rising or setting, lines to be drawn from her centre to the spectator and to the centre of the earth; these would form a right-angled triangle with the terrestrial radius, which is of a known length; and as the parallax or angle at the moon can be measured, all the angles and one side are given; whence the distance of the moon from the centre of the earth may be computed. The parallax of an object may be found, if two observers under the same meridian, but at a very great distance from one another, observe its zenith distance on the same day at the time of its passage over the meridian. By such contemporaneous observations at the Cape of Good Hope and at Berlin, the mean horizontal parallax of the moon was found to be 3459″, whence the mean distance of the moon is about sixty times the mean terrestrial radius, or 237360 miles nearly. Since the parallax is equal to the radius of the earth divided by the distance of the moon, it varies with the distance of the moon from the earth under the same parallel of latitude, and proves the ellipticity of the lunar orbit; when the moon is at her mean distance, it varies with the terrestrial radii, thus showing that the earth is not a sphere.

Although the method described is sufficiently accurate for finding the parallax of an object as near as the moon, it will not answer for the sun, which is so remote that the smallest error in observation would lead to a false result; but that difficulty is obviated by the transits of Venus. When that planet is in her nodes, or within $1\frac{1}{4}^{\circ}$ of them, that is, in, or nearly in, the plane of the ecliptic, she is occasionally seen to pass over the sun like a black spot. If we could imagine that the sun and Venus had no parallax, the line described by the planet on his disc and the duration of the transit would be the same to all the inhabitants of the earth; but as the semi-diameter of the earth has a sensible magnitude when viewed from the centre of the sun, the line described by the planet in its passage over his disc appears to be nearer to his centre, or farther from it, according to the position of the observer; so that the duration of the transit varies with the different points of the earth's surface at which it is observed. This difference of time, being entirely the effect of parallax, furnishes the means of computing it from the known motions of the earth and Venus, by the same method as for the eclipses of the sun. In fact, the ratio of the distances of Venus and the sun from the earth at the time of the transit are known from the theory of their elliptical motion, consequently the ratio of the parallaxes of these two bodies, being inversely as their distances, is given; and as the transit gives the difference of the parallaxes, that of the sun is obtained. In 1769, the parallax of the sun was determined by observations of a transit of Venus made at Wardhus in Lapland, and at Otaheite in the South Sea; the latter observation was the object of Cook's first voyage. The transit lasted about six hours at Otaheite, and the difference in duration at these two stations was eight minutes; whence the sun's horizontal parallax was found to be 8″·72: but by other considerations it has been reduced to 8″·577; from which the mean distance of the sun appears to be about 92246700 miles, or ninety-two millions of miles nearly. This is confirmed by an inequality in the motion of the moon, which depends upon the parallax of the sun, and which, when compared with observation, gives 8″·6 for the sun's parallax.

The parallax of Venus is determined by her transits, that of Mars by direct observation, and it is found to be nearly double that of the sun when the planet is in opposition. The distances of these two planets from the earth are therefore known in terrestrial radii; consequently their mean distances from the sun may be computed; and as the ratios of the distances of the planets from the sun are known by Kepler's law, their absolute distances in miles are easily found.

Far as the earth seems to be from the sun, it is near to him when compared with Uranus; that planet is no less than 1843000000 of miles from the luminary that warms and enlivens the world; situate on the verge of the system, the sun must appear to it not much larger than Venus does to us. The earth cannot even be visible as a telescopic object to a body so remote; yet man, the inhabitant of the earth, soars beyond the vast dimensions of the system to which his planet belongs, and assumes the diameter of its orbit as the base of a triangle, whose apex extends to the stars.

Sublime as the idea is, this assumption proves ineffectual, for the apparent places of the fixed stars are not sensibly changed by the earth's annual revolution; and with the aid derived from the refinements of modern astronomy, and of the most perfect of instruments, it is still a matter of doubt whether a sensible parallax has been detected even in the nearest of these remote suns. If a fixed star had the parallax of one second, its distance from the sun would be 20500000000000 of miles. At such a distance not only the terrestrial orbit shrinks to a point, but the whole solar system seen in the focus of the most powerful telescope, might be covered by the thickness of a spider's thread. Light flying at the rate of 200000 miles in a second, would take three years and seven days to travel over that space; one of the nearest stars may therefore have been kindled or extinguished more than three years before we could have been aware of so mighty an event. But this distance must be small when compared with that of the most remote of the bodies which are visible in the heavens. The fixed stars are undoubtedly luminous like the sun; it is therefore probable that they are not nearer to one another than the sun is to the nearest of them. In the milky way and the other starry nebulæ, some of the stars that seem to us to be close to others, may be far behind them in the boundless depth of space; nay, may be rationally supposed to be situate many thousand times farther off; light would therefore require thousands of years to come to the earth from those myriads of suns, of which our own is but 'the dim and remote companion.'

SECTION IX.

THE masses of such planets as have no satellites are known by comparing the inequalities they produce in the motions of the earth and of each other, determined theoretically, with the same inequalities given by observation, for the disturbing cause must necessarily be proportional to the effect it produces. But as the quantities of matter in any two primary planets are directly as the cubes of the mean distances at which their satellites revolve, and inversely as the squares of their periodic times, the mass of the sun and of any planets which have satellites may be compared with the mass of the earth. In this manner it is computed that the mass of the sun is 354936 times that of the earth; whence the great perturbations of the moon, and the rapid motion of the perigee and nodes of her orbit. Even Jupiter, the largest of the planets, is 1050 times less than the sun; or, as was recently determined by the perturbations of Encke's comet, appears the 1053·924th part of the sun. The mass of the moon is determined from several sources,—from her action on the terrestrial equator, which occasions the nutation in the axis of rotation; from her horizontal parallax; from an inequality she produces in the sun's longitude, and from her action on the tides. The three first quantities, computed from theory and compared with their observed values, give her mass respectively equal to the $\frac{1}{71}$, $\frac{1}{74 \cdot 2}$, and $\frac{1}{69 \cdot 2}$ part of that of the earth, which do not differ much from each other. Dr. Brinkley, Bishop of Cloyne, has found it to be $\frac{1}{80}$ from the constant of lunar nutation; but from the moon's action in raising the tides, her mass appears to be about the seventy-fifth part of that of the earth, a value that cannot differ much from the truth.

The apparent diameters of the sun, moon, and planets are determined by measurement; therefore their real diameters may be compared with that of the earth; for the real diameter of a planet is to the real diameter of the earth, or 7912 miles, as the apparent diameter of the planet to the apparent diameter of the earth as seen from the planet, that is, to twice the parallax of the planet. The mean apparent diameter of the sun is 1922″·8, and with the solar parallax 8″·577, it will be found that the diameter of the sun is about 886860 miles; therefore if the centre of the sun were to coincide with the centre of the earth, his volume would not only include the orbit of the moon, but would extend nearly as far again, for the moon's mean distance from the earth is about sixty times the earth's mean radius, or 237360 miles: so that twice the distance of the moon is 474960 miles, which differs but little from the solar radius; his equatorial radius is probably not much less than the major axis of the lunar orbit. The diameter of the moon is only 2160 miles; and Jupiter's diameter of 91509 miles is very much less than that of the sun; the diameter of Pallas does not much exceed 79 miles, so that an inhabitant of that planet, in one of our steam-carriages, might go round his world in a few hours.

Before entering on the theory of rotation, it may not be thought foreign to the subject to give some idea of the methods of computing the places of the planets, and of forming astronomical tables. Astronomy is now divided into the three distinct departments of theory, observation, and computation. Since the problem of the three bodies can only be solved by approximation, the analytical astronomer determines the position of a planet in space by a series of corrections. Its place in its circular orbit is first found, then the addition or subtraction of the equation of the centre to or from its mean place gives its position in the ellipse; this again is corrected by the application of the principal periodic inequalities; but as these are determined for some particular position of the three bodies, they require to be corrected to suit other relative positions This process is continued till the corrections become less than the errors of observation, when it is obviously unnecessary to carry the approximation further. By a similar method, the true latitude and distance of the planet from the sun are obtained.

All these quantities are given in terms of the time by general analytical formulæ; they will consequently give the exact place of the body in the heavens, for any time assumed at pleasure, provided they can be reduced to numbers; but before the calculator begins his task, the observer must furnish the necessary data. These are obviously the forms of the orbits, and their positions with regard to the plane of the ecliptic. It is therefore necessary to determine by observation for each planet, the length of the major axis of its orbit, the excentricity, the inclination of the orbit to the plane of the ecliptic, the longitudes of its perihelion and ascending node at a given time, the periodic time of the planet, and its longitude at any instant, arbitrarily assumed as an origin from whence all its subsequent and antecedent longitudes are estimated. Each of these quantities is determined from that position of the planet on which it has most influence. For example, the sum of the greatest and least distances of the planet from the sun is equal to the major axis of the orbit, and their difference is equal to the excentricity; the longitude of the planet, when at its least distance from the sun, is the same with the longitude of the perihelion; the greatest latitude of the planet is equal to the inclination of the orbit; and the longitude of the planet, when in the plane of the ecliptic in passing towards the north, is the longitude of the ascending node. But, notwithstanding the excellence of instruments and the accuracy of modern observers, the unavoidable errors of observation can only be compensated by finding the value of each element from the mean of perhaps a thousand, or even many thousands of observations: for as it is probable that the errors are not all in one direction, but that some are in excess and others in defect, they will compensate each other when combined.

However, the values of the elements determined separately can only be regarded as approximate, because they are so connected that the estimation of any one independently will induce errors in the others, for the excentricity depends upon the longitude of the perihelion, the mean motion depends upon the major axis, the longitude of the node upon the inclination of the orbit, and *vice versâ*, consequently the place of a planet computed with the approximate data, will differ from its observed place: then the difficulty is to ascertain what elements are most in fault, since the difference in question is the error of all, but that is obviated by finding the errors of some thousands of observations, and combining them so as to correct the elements simultaneously, and to make the sum of the squares of the errors a minimum with regard to each element. The method of accomplishing this depends upon the Theory of Probabilities, a subject fertile in most important results in the various departments of science and of civil life, and quite indispensable in the determination of astronomical data. A series of observations continued for some years will give approximate values of the secular and periodic inequalities, which must be corrected from time to time till theory and observation agree; and when all these quantities are determined in numbers, the longitude, latitude, and distances of the planet from the sun are computed for stated intervals, and formed into tables, arranged according to the time estimated from a given epoch, so that the place of the body may be determined from them by inspection alone, at any instant, for perhaps a thousand years before and after that epoch. By this tedious process, tables have been computed for eleven planets, besides the moon and the satellites of Jupiter. Those of the four new planets are astonishingly perfect, considering that these bodies have not been discovered more than thirty years, and a much longer time is requisite to develop their inequalities.

SECTION X.

The oblate form of several of the planets indicates rotatory motion; this has been confirmed, in most cases, by tracing spots on their surface, by which their poles and times of rotation have been determined. The rotation of Mercury is unknown, on account of his proximity to the sun; and that of the new planets has not yet been ascertained. The sun revolves in twenty-five days and ten hours about an axis which is directed towards a point half-way between the pole-star and Lyra, the plane of rotation being inclined by 7° 20′, or a little more than seven degrees, to the plane of the ecliptic. From the rotation of the sun, there is every reason to believe that he has a progressive motion in space, although the direction to which he tends is unknown: but in consequence of the reaction of the planets, he describes a small irregular orbit about the centre of inertia of the system, never deviating from his position by more than twice his own diameter, or a little more than seven times the distance of the moon from the earth. The sun and all his attendants rotate from west to east, on axes that remain nearly parallel to themselves in every point of their orbit, and with angular velocities that are sensibly uniform. Although the uniformity in the direction of their rotation is a circumstance hitherto unaccounted for in the economy of nature, yet from the design and adaptation of every other part to the perfection of the whole, a coincidence so remarkable cannot be accidental; and as the revolutions of the planets and satellites are also from west to east, it is evident that both must have arisen fron the primitive cause which has determined the planetary motions. Indeed, La Place has computed the probability to be as four millions to one, that all the motions of the planets, both of rotation and revolution, were at once imparted by an original common cause, but of which we know neither the nature nor the epoch.

The larger planets rotate in shorter periods than the smaller planets and the earth, their compression is consequently greater, and the action of the sun and of their satellites occasions a nutation in their axes, and a precession of their equinoxes similar to that which obtains in the terrestrial spheroid, from the attraction of the sun and moon on the prominent matter at the equator. It is an evident consequence of Kepler's law of the squares of the periodic times of the planets being as the cubes of the major axes of their orbits, that the heavenly bodies move slower the farther they are from the sun. In comparing the periods of the revolutions of Jupiter and Saturn with the times of their rotation, it appears that a year of Jupiter contains nearly ten thousand of his days, and that of Saturn about thirty thousand Saturnian days.

The appearance of Saturn is unparalleled in the system of the world; he is a spheroid about 900 times larger than the earth, surrounded by a ring even brighter than himself, which always remains suspended in the plane of his equator, and viewed with a very good telescope, it is found to consist of two concentric rings, divided by a dark band. The mean distance of the interior part of this double ring from the surface of the planet is about 22240 miles, it is no less than 33360 miles broad, but, by estimation, its thickness does not much exceed 274 miles, so that it appears like a plane. By the laws of mechanics, it is impossible that this body can retain its position by the adhesion of its particles alone; it must necessarily revolve with a velocity that will generate a centrifugal force sufficient to balance the attraction of Saturn. Observation confirms the truth of these principles, showing that the rings rotate about the planet in ten hours and a half, which is considerably less than the time a satellite would take to revolve about Saturn at the same distance. Their plane is inclined to the ecliptic, at an angle of 28° 39′ 45″; and, in consequence of this obliquity of position, they always appear elliptical to us, but with an excentricity so variable as even to be occasionally like a straight line drawn across the planet. In the beginning of October, 1832, the plane of the rings passed through the centre of the earth; in that position they are only visible with very superior instruments, and appear like a fine line across the disc of Saturn. About the middle of December, in the same year, the rings became invisible, with ordinary instruments, on account of their plane passing through the sun. In the end of April, 1833, the rings vanished a second time, and reappeared in June of that year.

It is a singular result of theory, that the rings could not maintain their stability of rotation if they were every where of uniform thickness;

for the smallest disturbance would destroy the equilibrium, which would become more and more deranged till, at last, they would be precipitated on the surface of the planet. The rings of Saturn must therefore be irregular solids of unequal breadth in different parts of the circumference, so that their centres of gravity do not coincide with the centres of their figures. Professor Struve has also discovered that the centre of the ring is not concentric with the centre of Saturn; the interval between the outer edge of the globe of the planet, and the outer edge of the ring on one side, is 11″·073, and, on the other side, the interval is 11″·288, consequently there is an excentricity of the globe in the ring of 0″·215. If the rings obeyed different forces, they would not remain in the same plane, but the powerful attraction of Saturn always maintains them and his satellites in the plane of his equator. The rings, by their mutual action, and that of the sun and satellites, must oscillate about the centre of Saturn, and produce phenomena of light and shadow whose periods extend to many years.

The periods of rotation of the moon and the other satellites are equal to the times of their revolutions, consequently these bodies always turn the same face to their primaries: however, as the mean motion of the moon is subject to a secular inequality, which will ultimately amount to many circumferences, if the rotation of the moon were perfectly uniform, and not affected by the same inequalities, it would cease exactly to counterbalance the motion of revolution; and the moon, in the course of ages, would successively and gradually discover every point of her surface to the earth. But theory proves that this never can happen; for the rotation of the moon, though it does not partake of the periodic inequalities of her revolution, is affected by the same secular variations, so that her motions of rotation and revolution round the earth will always balance each other, and remain equal. This circumstance arises from the form of the lunar spheroid, which has three principal axes of different lengths at right angles to each other.

The moon is flattened at her poles from her centrifugal force, therefore her polar axis is the least; the other two are in the plane of her equator, but that directed towards the earth is the greatest. The attraction of the earth, as if it had drawn out that part of the moon's equator, constantly brings the greatest axis, and consequently the same hemisphere, towards us, which makes her rotation participate in the secular variations in her mean motion of revolution. Even if the angular velocities of rotation and revolution had not been nicely balanced in the beginning of the moon's motion, the attraction of the earth would have recalled the greatest axis to the direction of the line joining the centres of the moon and earth; so that it would have vibrated on each side of that line in the same manner as a pendulum oscillates on each side of the vertical from the influence of gravitation. No such libration is perceptible; and as the smallest disturbance would make it evident, it is clear that if the moon has ever been touched by a comet, the mass of the latter must have been extremely small; for if it had been only the hundred thousandth part of that of the earth, it would have rendered the libration sensible. According to analysis, a similar libration exists in the motions of Jupiter's satellites, which still remains insensible to observation.

It is true the moon is liable to librations depending upon the position of the spectator; at her rising, part of the western edge of her disc is visible, which is invisible at her setting, and the contrary takes place with regard to her eastern edge. There are also librations arising from the relative positions of the earth and moon in their respective orbits, but as they are only optical appearances, one hemisphere will be eternally concealed from the earth. For the same reason, the earth, which must be so splendid an object to one lunar hemisphere, will be for ever veiled from the other. On account of these circumstances, the remoter hemisphere of the moon has its day a fortnight long, and a night of the same duration, not even enlightened by a moon, while the favoured side is illuminated by the reflection of the earth during its long night. A planet exhibiting a surface thirteen times larger than that of the moon, with all the varieties of clouds, land, and water coming successively into view, would be a splendid object to a lunar traveller in a journey to his antipodes. The great height of the lunar mountains probably has a considerable influence on the phenomena of her motion, the more so as her compression is small, and her mass considerable. In the curve passing through the poles, and that diameter of the moon which always points to the earth, nature has furnished a permanent meridian, to which the different spots on her surface have been referred, and their positions determined with as much accuracy as those of many of the most remarkable places on the surface of our globe.

The distance and minuteness of Jupiter's satellites render it extremely difficult to ascertain their rotation. It was, however, accomplished by Sir William Herschel from their relative brightness. He observed that they alternately exceed each other in brilliancy, and, by comparing the maxima and minima of their illumination with their positions relatively to the sun and to their primary, he found that, like the moon, the time of their rotation is equal to the period of their revolution about Jupiter. Miraldi was led to the same conclusion with regard to the fourth satellite, from the motion of a spot on its surface.

SECTION XI.

The rotation of the earth, which determines the length of the day, may be regarded as one of the most important elements in the system of the world. It serves as a measure of time, and forms the standard of comparison for the revolutions of the celestial bodies, which, by their proportional increase or decrease, would soon disclose any changes it might sustain. Theory and observation concur in proving that, among the innumerable vicissitudes which prevail throughout creation, the period of the earth's diurnal rotation is immutable. A fluid, falling from a higher to a lower level, carries with it the velocity due to its revolution with the earth at a greater distance from the centre; it will therefore accelerate, although to an almost infinitesimal extent, the earth's daily rotation. The sum of all these increments of velocity, arising from the descent of all the rivers on the earth's surface, would in time become perceptible, did not nature, by the process of evaporation, raise the waters back to their sources; and thus, by again removing matter to a greater distance from the centre, destroy the velocity generated by its previous approach; so that the descent of rivers does not affect the earth's rotation. Enormous masses projected by volcanos from the equator to the poles, and the contrary, would indeed affect it, but there is no evidence of such convulsions. The disturbing action of the moon and planets, which has so powerful an effect on the revolution of the earth, in no way influences its rotation; the constant friction of the trade-winds on the mountains and continents between the tropics does not impede its velocity, which theory even proves to be the same as if the sea, together with the earth, formed one solid mass. But although these circumstances be inefficient, a variation in the mean temperature would certainly occasion a corresponding change in the velocity of rotation; for, in the science of dynamics, it is a principle in a system of bodies, or of particles revolving about a fixed centre, that the momentum, or sum of the products of the mass of each, into its angular velocity and distance from the centre, is a constant quantity, if the system be not deranged by a foreign cause. Now, since the number of particles in the system is the same, whatever its temperature may be, when their distances from the centre are diminished, their angular velocity must be increased, in order that the preceding quantity may still remain constant. It follows then, that, as the primitive momentum of rotation with which the earth was projected into space must necessarily remain the same, the smallest decrease in heat, by contracting the terrestrial spheroid, would accelerate its rotation, and consequently diminish the length of the day. Notwithstanding the constant accession of heat from the sun's rays, geologists have been induced to believe, from the fossil remains, that the mean temperature of the globe is decreasing.

The high temperature of mines, hot springs, and, above all, the internal fires which have produced and do still occasion such devastation on our planet, indicate an augmentation of heat towards its centre; the increase of density, corresponding to the depth and the form of the spheroid, being what theory assigns to a fluid mass in rotation, concur to induce the idea that the temperature of the earth was originally so high as to reduce all the substances of which it is composed to a state of fusion, or of vapour, and that, in the course of ages, it has cooled down to its present state; that it is still becoming colder, and that it will continue to do so till the whole mass arrives at the temperature of the medium in which it is placed, or rather at a state of equilibrium between this temperature, the cooling power of its own radiation, and the heating effect of the sun's rays.

Previous to the formation of ice at the poles, the ancient lands of our northern latitudes, long since obliterated, might, no doubt, have been capable of producing those tropical plants whose debris, swept into the deep at these remote periods, are preserved in the coal measures which must have been formed in the abysses of the ocean prior to the elevation of the modern continents and islands above its surface. But, even if the decreasing temperature of the earth be sufficient to produce the observed effects, it must be extremely slow in its operation; for, in consequence of the rotation of the earth being a measure of the periods of the celestial motions, it has been proved that, if the length of the day had decreased by the three-thousandth part of a second since the observations of Hipparchus, two thousand years ago, it would have diminished the secular equation of the moon by $4''\cdot4$. It is therefore beyond a doubt that the mean temperature of the earth cannot have sensibly varied during that time; if, then, the appearances exhibited by the strata are really owing to a decrease of internal temperature, it either shows the immense periods requisite to produce geological changes, to which two thousand years are as nothing, or that the mean temperature of the earth had arrived at a state of equilibrium before these observations.

However strong the indications of the primitive fluidity of the earth, as there is no direct proof of it, the hypothesis can only be regarded as very probable; but one of the most profound philosophers and elegant writers of modern times has found in the secular variation in the excentricity of the terrestrial orbit an evident cause of decreasing temperature. That accomplished author, in pointing out the mutual dependences of phenomena, says, 'It is evident that the mean temperature of the whole surface of the globe, in so far as it is maintained by the action of the sun at a higher degree than it would have were the sun extinguished, must depend on the mean quantity of the sun's rays which it receives, or—which comes to the same thing—on the total quantity received in a given invariable time; and the length of the year being unchangeable in all the fluctuations of the planetary system, it follows that the total amount of solar radiation will determine, *cæteris paribus*, the general climate of the earth. Now, it is not difficult to show that this amount is inversely proportional to the minor axis of the ellipse described by the earth about the sun, regarded as slowly variable; and that, therefore, the major axis remaining, as we know it to be, constant, and the orbit being actually in a state of approach to a circle, and consequently the minor axis being on the increase, the mean annual amount of solar radiation received by the whole earth must be actually on the decrease. We have therefore an evident real cause to account for the phenomenon.' The limits of the variation in the excentricity of the earth's orbit are unknown; but if its ellipticity has ever been as great as that of the orbit of Mercury or Pallas, the mean temperature of the earth must have been sensibly higher than it is at present; whether it was great enough to render our northern climates fit for the production of tropical plants, and for the residence of the elephant and other animals now inhabitants of the torrid zone, it is impossible to say.

The relative quantity of heat received by the earth at different moments during a single revolution varies with the position of the perigee, which accomplishes a tropical revolution in 21067 years. In the year 1245 of our era, and 19822 years before it, the perigee coincided with the winter solstice; at both these periods the earth was nearer the sun during the summer, and farther from him in the winter, than in any other position of the apsides; the extremes of temperature must therefore have been greater than at present; but as the terrestrial orbit was probably more elliptical at the distant epoch, the heat of the summers must have been very great, though possibly compensated by the rigour of the winters; at all events, none of these changes affect the length of the day.

It appears, from the marine shells found on the tops of the highest mountains, and in almost every part of the globe, that immense continents have been elevated above the ocean, which must have engulphed others. Such a catastrophe would be occasioned by a variation in the position of the axis of rotation on the surface of the earth; for the seas, tending to a new equator, would leave some portions of the globe and overwhelm others. Now, it is found by the laws of mechanics that, in every body, be its form or density what it may, there are at least three axes at right angles to each other, round any one of which, if the solid begins to rotate, it will continue to revolve for ever, provided it be not disturbed by a foreign cause, but that the rotation about any other axis will only be permanent for an instant; consequently the poles or extremities of the instantaneous axis of rotation would perpetually change their position on the surface of the body. In an ellipsoid of revolution, the polar diameter, and every diameter in the plane of the equator, are the only permanent axes of rotation; consequently, if the ellipsoid were to begin to revolve about any diameter between the pole and the equator, the motion would be so unstable, that the axis of rotation and the position of the poles would change every instant. Hence, as the earth does not differ much from this figure, if it did not turn round one of its principal axes, the position of the poles would change daily; the equator, which is 90° distant, would undergo corresponding variations; and the geographical latitudes of all places, being estimated from the equator, assumed to be fixed, would be perpetually changing.

A displacement in the position of the poles of only two hundred miles would be sufficient to produce these effects, and would immediately be detected; but as the latitudes are found to be invariable, it may be concluded that the terrestrial spheroid must have revolved about the same axis for ages. The earth and planets differ so little from ellipsoids of revolution, that, in all probability, any libration from one axis to another, produced by the primitive impulse which put them in motion, must have ceased soon after their creation from the friction of the fluids at their surfaces.

Theory also proves that neither nutation, precession, nor any of the disturbing forces that affect the system, have the smallest influence on the axis of rotation, which maintains a permanent position on the surface, if the earth be not disturbed in its rotation by a foreign cause, as the collision of a comet, which might have happened in the immensity of time. But had that been the case, its effects would still have been perceptible in the variations of the geographical latitudes. If we suppose that such an event had taken place, and that the disturbance had been very great, equilibrium could then only have been restored, with regard to a new axis of rotation, by the rushing of the seas to the new equator, which they must have continued to do till the surface was everywhere perpendicular to the direction of gravity. But it is probable that such an accumulation of the waters would not be sufficient to restore equilibrium if the derangement had been great, for the mean density of the sea is only about a fifth part of the mean density of the earth, and the mean depth of the Pacific Ocean is not more than four miles, whereas the equatorial radius of the earth exceeds the polar radius by about twenty-five miles; consequently, the influence of the sea on the direction of gravity is very small; and as it thus appears that a great change in the position of the axis is incompatible with the law of equilibrium, the geological phenomena in question must be ascribed to an internal cause. Indeed, it is now demonstrated that the strata containing marine diluvia, which are in lofty situations, must have been formed at the bottom of the ocean, and afterwards upheaved by the action of subterraneous fires. Besides, it is clear, from the mensuration of the arcs of the meridian, and the length of the seconds pendulum, as well as from the lunar theory, that the internal strata, and also the external outline of the globe, are elliptical, their centres being coincident, and their axes identical, with that of the surface,—a state of things which, according to the distinguished author lately quoted, is incompatible with a subsequent accommodation of the surface to a new and different state of rotation from that which determined the original distribution of the component matter. Thus, amidst the mighty revolutions which have swept innumerable races of organized beings from the earth, which have elevated plains, and buried mountains in the ocean, the rotation of the earth, and the position of the axes on its surface, have undergone but slight variations.

It not only appears that the strata of the terrestrial spheroid are concentric and elliptical, but the lunar inequalities show that they increase in density from the surface of the earth to its centre. This would certainly have happened if the earth had originally been fluid, for the denser parts must have subsided towards the centre, as it approached a state of equilibrium; but the enormous pressure of the superincumbent mass is a sufficient cause for the phenomenon. Professor Leslie observes that air, compressed into the fiftieth part of its volume, has its elasticity fifty times augmented; if it continue to contract at that rate, it would, from its own incumbent weight, acquire the density of water at the depth of thirty-four miles. But water itself would have its density doubled at the depth of ninety-three miles, and would even attain the density of quicksilver at a depth of 362 miles. In descending, therefore, towards the centre, through nearly 4000 miles, the condensation of ordinary substances would surpass the utmost powers of conception. Dr. Young says that steel would be compressed into one-fourth and stone into one-eighth of its bulk at the earth's centre. However, we are yet ignorant of the laws of compression of solid bodies beyond a certain limit; though, from the experiments of Mr. Perkins, they appear to be capable of a greater degree of compression than has generally been imagined.

But a density so extreme is not borne out by astronomical observation. It might seem to follow, therefore, that our planet must have a widely cavernous structure, and that we tread on a crust or shell whose thickness bears a very small proportion to the diameter of its sphere. Possibly, too, this great condensation at the central regions may be counterbalanced by the increased elasticity due to a very elevated temperature.

SECTION XII.

It has been shown that the axis of rotation is invariable on the surface of the earth, and observation, as well as theory, prove that, were it not for the action of the sun and moon on the matter at the equator, it would remain parallel to itself in every point of its orbit.

The attraction of an external body not only draws a spheroid towards it, but, as the force varies inversely as the square of the distance, it gives it a motion about its centre of gravity, unless when the attracting body is situate in the prolongation of one of the axes of the spheroid. The plane of the equator is inclined to the plane of the ecliptic at an angle of $23^\circ\ 27'\ 36''\cdot7$; and the inclination of the lunar orbit on the same is $5^\circ\ 8'\ 47''\cdot9$; consequently, from the oblate figure of the earth, the sun and moon acting obliquely and unequally on the different parts of the terrestrial spheroid, urge the plane of the equator from its direction, and force it to move from east to west, so that the equinoctial points have a slow retrograde motion on the plane of the ecliptic of 50″·37572 annually. The direct tendency of this action is to make the planes of the equator and ecliptic coincide, but it is balanced by the tendency of the earth to return to stable rotation about the polar diameter, which is one of its principal axes of rotation; therefore the inclination of the two planes remains constant, as a top spinning preserves the same inclination to the plane of the horizon. Were the earth spherical, this effect would not be produced, and the equinoxes would always correspond with the same points of the ecliptic, at least as far as this kind of motion is concerned. But another and totally different cause which operates on this motion has already been mentioned. The action of the planets on one another and on the sun occasions a very slow variation in the position of the plane of the ecliptic, which affects its inclination to the plane of the equator, and gives the equinoctial points a slow but direct motion on the ecliptic of 0″·15272 annually, which is entirely independent of the figure of the earth, and would be the same if it were a sphere. Thus the sun and moon, by moving the plane of the equator, cause the equinoctial points to retrograde on the ecliptic, and the planets, by moving the plane of the ecliptic, give them a direct motion, though much less than the former; consequently, the difference of the two is the mean precession, which is proved, both by theory and observation, to be about 50″·223 annually.

As the longitudes of all the fixed stars are increased by this quantity, the effects of precession are soon detected; it was accordingly discovered by Hipparchus, in the year 128 before Christ, from a comparison of his own observations with those of Timocharis, 155 years before. In the time of Hipparchus, the entrance of the sun into the constellation Aries was the beginning of spring, but since that time the equinoctial points have receded 30°, so that the constellations called the signs of the zodiac are now at a considerable distance from those divisions of the ecliptic which bear their names. Moving at the rate of 50″·223 annually, the equinoctial points will accomplish a revolution in 25805 years; but as the precession varies in different centuries, the extent of this period will be slightly modified. Since the motion of the sun is direct, and that of the equinoctial points retrograde, he takes a shorter time to return to the equator than to arrive at the same stars; so that the tropical year of 365·242219 mean solar days must be increased by the time he takes to move through an arc of 50″·223, in order to have the length of the sidereal year. By simple proportion, it is the 0·014154th part of a day, so that the sidereal year contains 365·256373 mean solar days.

The mean annual precession is subject to a secular variation; for, although the change in the plane of the ecliptic, in which the orbit of the sun lies, be independent of the form of the earth, yet, by bringing the sun, moon, and earth into different relative positions, from age to age, it alters the direct action of the two first on the prominent matter at the equator: on this account, the motion of the equinox is greater by 0″·455 now than it was in the time of Hipparchus; consequently, the actual length of the tropical year is about 4ˢ·21 shorter than it was at that time. The utmost change that it can experience from this cause amounts to 43 seconds.

Such is the secular motion of the equinoxes; but it is sometimes increased and sometimes diminished by periodic variations, whose periods depend upon the relative positions of the sun and moon with regard to the earth, and which are occasioned by the direct action of these bodies on the equator. Dr. Bradley discovered that by this action the moon causes the pole of the equator to describe a small ellipse in the heavens, the diameters of which are 16″ and 20″. The period of this inequality is 19 years, the time employed by the nodes of the lunar orbit to accomplish a revolution. The sun causes a small variation in the description of this ellipse; it runs through its period in half a year. This nutation in the earth's axis affects both the precession and obliquity with small periodic variations; but, in consequence of the secular variation in the position of the terrestrial orbit, which is chiefly owing to the disturbing energy of Jupiter on the earth, the obliquity of the ecliptic is annually diminished by $0''\cdot445$, or, according to Bessel, by $0''\cdot457$. This variation in the course of ages may amount to ten or eleven degrees; but the obliquity of the ecliptic to the equator can never vary more than $2^\circ 42'$ or 3°, since the equator will follow in some measure the motion of the ecliptic.

It is evident that the places of all the celestial bodies are affected by precession and nutation, and therefore all observations of them must be corrected for these inequalities.

The densities of bodies are proportional to their masses divided by their volumes; hence, if the sun and planets be assumed to be spheres, their volumes will be as the cubes of their diameters. Now, the apparent diameters of the sun and earth, at their mean distance, are $1922''\cdot8$ and $17''\cdot154$, and the mass of the earth is the 354936th part of that of the sun taken as the unit: it follows, therefore, that the earth is nearly four times as dense as the sun; but the sun is so large, that his attractive force would cause bodies to fall through about 334·65 feet in a second; consequently, if he were habitable by human beings, they would be unable to move, since their weight would be thirty times as great as it is here. A man of moderate size would weigh about two tons at the surface of the sun, whereas, at the surface of the four new planets, he would be so light, that it would be impossible to stand steady, since he would only weigh a few pounds. All the planets and satellites appear to be of less density than the earth. The motions of Jupiter's satellites show that his density increases towards his centre: were his mass homogeneous, his equatorial and polar axes would be in the ratio of 41 to 36, whereas they are observed to be only as 41 to 38. The singular irregularities in the form of Saturn, and the great compression of Mars, prove the internal structure of these two planets to be very far from uniform.

SECTION XIII.

ASTRONOMY has been of immediate and essential use in affording invariable standards for measuring duration, distance, magnitude, and velocity. The sidereal day, measured by the time elapsed between two consecutive transits of any star at the same meridian, and the sidereal year, are immutable units with which all great periods of time are compared; the oscillations of the isochronous pendulum measure its smaller portions. By these invariable standards alone, we can judge of the slow changes that other elements of the system may have undergone in the lapse of ages.

The returns of the sun to the meridian, and to the same equinox or solstice, have been universally adopted as the measure of our civil days and years. The solar or astronomical day is the time that elapses between two consecutive noons or midnights; it is consequently longer than the sidereal day, on account of the proper motion of the sun during a revolution of the celestial sphere; but, as the sun moves with greater rapidity at the winter than at the summer solstice, the astronomical day is more nearly equal to the sidereal day in summer than in winter. The obliquity of the ecliptic also affects its duration, for in the equinoxes the arc of the equator is less than the corresponding arc of the ecliptic, and in the solstices it is greater. The astronomical day is therefore diminished in the first case, and increased in the second. If the sun moved uniformly in the equator at the rate of $59' \, 8'' \cdot 3$ every day, the solar days would be all equal; the time, therefore, which is reckoned by the arrival of an imaginary sun at the meridian, or of one which is supposed to move uniformly in the equator, is denominated mean solar time, such as is given by clocks and watches in common life: when it is reckoned by the arrival of the real sun at the meridian, it is apparent time, such as is given by dials. The difference between the time shown by a clock and a dial is the equation of time given in the Nautical Almanac, sometimes amounting to as much as sixteen minutes. The apparent and mean time coincide four times in the year.

The astronomical day begins at noon, but in common reckoning the day begins at midnight. In England it is divided into twenty-four hours, which are counted by twelve and twelve; but in France, astronomers, adopting the decimal division, divide the day into ten hours, the hour into one hundred minutes, and the minute into a hundred seconds, because of the facility in computation, and in conformity with their system of weights and measures. This subdivision is not used in common life, nor has it been adopted in any other country; and although some scientific writers in France still employ that division of time, the custom is beginning to wear out. The mean length of the day, though accurately determined, is not sufficient for the purposes either of astronomy or civil life. The tropical or civil year of 365·242219 mean solar days, the time elapsed between the consecutive returns of the sun to the mean equinoxes or solstices, including all the changes of the seasons, is a natural cycle peculiarly suited for a measure of duration. It is estimated from the winter solstice, the middle of the long annual night under the poles. But although the length of the civil year is pointed out by nature as a measure of long periods, the incommensurability that exists between the length of the day and the revolution of the sun renders it difficult to adjust the estimation of both in whole numbers. If the revolution of the sun were accomplished in 365 days, all the years would be of precisely the same number of days, and would begin and end with the sun at the same point of the ecliptic; but as the sun's revolution includes the fraction of a day, a civil year and a revolution of the sun have not the same duration. Since the fraction is nearly the fourth of a day, in four years it is nearly equal to a revolution of the sun, so that the addition of a supernumerary day every fourth year nearly compensates the difference; but, in process of time, further correction will be necessary, because the fraction is less than the fourth of a day. In fact, if a bissextile be suppressed at the end of three out of four centuries, the year so determined will only exceed the true year by an extremely small fraction of a day;

and if, in addition to this, a bissextile be suppressed every 4000 years, the length of the year will be nearly equal to that given by observation. Were the fraction neglected, the beginning of the year would precede that of the tropical year, so that it would retrograde through the different seasons in a period of about 1507 years. The Egyptians estimated the year at 365·25 days, by which they lost one year in every 14601—their Sothiac period. The division of the year into months is very old and almost universal; but the period of seven days, by far the most permanent division of time, and the most ancient monument of astronomical knowledge, was used by the Brahmins in India with the same denominations employed by us, and was alike found in the calendars of the Jews, Egyptians, Arabs, and Assyrians; it has survived the fall of empires, and has existed among all successive generations, a proof of their common origin.

The new moon immediately following the winter solstice in the 707th year of Rome was made the 1st of January of the first year of Julius Cæsar; the 25th of December of his forty-fifth year is considered as the date of Christ's nativity; and Cæsar's forty-sixth year is assumed to be the first of our era. The preceding year is called the first year before Christ by chronologists, but by astronomers it is called the year 0. The astronomical year begins on the 31st of December, at noon; and the date of an observation expresses the days and hours which have actually elapsed since that time.

Some remarkable astronomical eras are determined by the position of the major axis of the solar ellipse, which depends upon the direct motion of the perigee and the precession of the equinoxes conjointly, the annual motion of the one being 11″·2936, and that of the other 50″·223; hence the axis, moving at the rate of 61″·5166 annually, accomplishes a tropical revolution in 21067 years. It coincided with the line of the equinoxes 4000 or 4022 years before the Christian era, much about the time chronologists assign for the creation of man. In 6512 the major axis will again coincide with the line of the equinoxes, but then the solar perigee will coincide with the equinox of spring, whereas at the creation of man it coincided with the autumnal equinox. In the year 1245, the major axis was perpendicular to the line of the equinoxes, then the solar perigee coincided with the solstice of winter, and the apogee with the solstice of summer. According to La Place, who computed these periods from different data, the last coincidence happened in the year 1250 of our era, which induced him to propose that year as a universal epoch, the vernal equinox of the year 1250 to be the first day of the first year.

The variation in the position of the solar ellipse occasions corresponding changes in the length of the seasons. In its present position, spring is shorter than summer, and autumn longer than winter; and while the solar perigee continues as it now is, between the solstice of winter and the equinox of spring, the period including spring and summer will be longer than that including autumn and winter. In this century the difference is between seven and eight days. The intervals will be equal towards the year 6512, when the perigee coincides with the equinox of spring, but when it passes that point, the spring and summer, taken together, will be shorter than the period including the autumn and winter. These changes will be accomplished in a tropical revolution of the major axis of the earth's orbit, which includes an interval of 21067 years; and as the seasons are opposed to each other in the northern and southern hemispheres, they alternately receive, for a period of 10534 years, a greater portion of light and heat. Were the orbit circular, the seasons would be equal; their difference arises from the excentricity of the orbit, small as it is; but the changes are so trifling, as to be imperceptible in the short space of human life.

No circumstance in the whole science of astronomy excites a deeper interest than its application to chronology. "Whole nations," says La Place, "have been swept from the earth, with their languages, arts, and sciences, leaving but confused masses of ruins to mark the place where mighty cities stood; their history, with the exception of a few doubtful traditions, has perished; but the perfection of their astronomical observations marks their high antiquity, fixes the periods of their existence, and proves that, even at that early time, they must have made considerable progress in science." The ancient state of the heavens may now be computed with great accuracy; and by comparing the results of computation with ancient observations, the exact period at which they were made may be verified if true, or, if false, their error may be detected. If the date be accurate, and the observation good, it will verify the accuracy of modern tables, and will show to how many centuries they may be extended, without the fear of error. A few examples will show the importance of the subject.

At the solstices the sun is at his greatest distance from the equator, consequently his declination at these times is equal to the obliquity of the ecliptic, which, in former times, was determined from the meridian length of the shadow of the stile of a dial on the day of the solstice. The lengths of the meridian shadow at the summer and winter solstice are recorded to have been observed at the city of Layang, in China, 1100 years before the Christian era. From these, the distances of the sun from the zenith of the city of Layang are known. Half the sum of these zenith distances determines the latitude, and half their difference gives the obliquity of the ecliptic at the period of the observation; and as the law of the variation of the obliquity is known, both the time and place of the observations have been verified by computations from modern tables. Thus the Chinese had made some advances in the science of astronomy at that early period; their whole chronology is founded on the observation of eclipses, which prove the existence of that empire for more than 4700 years. The epoch of the lunar tables of the Indians, supposed by Bailly to be 3000 years before the Christian era, was proved by La Place, from the acceleration of the moon, not to be more ancient than the time of Ptolemy, who lived in the second century after it. The great inequality of Jupiter and Saturn, whose cycle embraces 929 years, is peculiarly fitted for marking the civilization of a people. The Indians had determined the mean motions of these two planets in that part of their periods when the apparent mean motion of Saturn was at the slowest, and that of Jupiter the most rapid. The periods in which that happened was 3102 years before the Christian era, and the year 1491 after it. The returns of comets to their perihelia may possibly mark the present state of astronomy to future ages.

The places of the fixed stars are affected by the precession of the equinoxes; and as the law of that variation is known, their positions at any time may be computed. Now Eudoxus, a contemporary of Plato, mentions a star situate in the pole of the equator, and it appears from computation, that κ Draconis was not very far from that place about 3000 years ago; but as it is only about 2150 years since Eudoxus lived, he must have described an anterior state of the heavens, supposed to be the same that was mentioned by Chiron, about the time of the siege of Troy. Every circumstance concurs in showing that astronomy was cultivated in the highest ages of antiquity.

It is possible that a knowledge of astronomy may lead to the interpretation of hieroglyphical characters. Astronomical signs are often found on the ancient Egyptian monuments, probably employed by the priests to record dates. The author had occasion to witness an instance of this most interesting application of astronomy, in ascertaining the date of a papyrus, sent from Egypt by Mr. Salt, in the hieroglyphical researches of the late Dr. Thomas Young, whose profound and varied acquirements do honour to his country and to the age in which he lived. The manuscript was found in a mummy-case; it proved to be a horoscope of the age of Ptolemy, and its antiquity was determined from the configuration of the heavens at the time of its construction.

The form of the earth furnishes a standard of weights and measures for the ordinary purposes of life, as well as for the determination of the masses and distances of the heavenly bodies. The length of the pendulum vibrating seconds of mean solar time, in the latitude of London, forms the standard of the British measure of extension. Its length oscillating in vacuo at the temperature of 62° of Fahrenheit, and reduced to the level of the sea was determined, by Captain Kater, to be 39·1392 inches. The weight of a cubic inch of water at the temperature of 62° of Fahrenheit, barometer 30 inches, was also determined in parts of the imperial troy pound, whence a standard both of weight and capacity is deduced. The French have adopted the metre equal to 3·2808992 English feet for their unit of linear measure, which is the ten-millionth part of that quadrant of the meridian passing through Formentera and Greenwich, the middle of which is nearly in the forty-fifth degree of latitude. Should the national standards of the two countries be lost in the vicissitude of human affairs, both may be recovered, since they are derived from natural standards presumed to be invariable. The length of the pendulum would be found again with more facility than the metre; but as no measure is mathematically exact, an error in the original standard may at length become sensible in measuring a great extent, whereas the error that must necessarily arise in measuring the quadrant of the meridian is rendered totally insensible by subdivisions, in taking its ten-millionth part. The French have adopted the decimal division, not only in time, but in their degrees, weights, and measures, on account of the very great facility it affords in computation. It has not been adopted by any other people, though nothing is more desirable than that all nations should concur in using the same division and standards, not only on account of convenience, but as affording a more definite idea of quantity. It is singular that the decimal division of the day, of degrees, weights, and measures, was employed in China 4000 years ago; and that at the time Ibn Junis made his observations at Cairo, about the year 1000 of the Christian era, the Arabs were in the habit of employing the vibrations of the pendulum in their astronomical observations as a measure of time.

SECTION XIV.

One of the most immediate and remarkable effects of a gravitating force external to the earth, is the alternate rise and fall of the surface of the sea twice in the course of a lunar day, or 24^h 50^m 48^s of mean solar time. As it depends upon the action of the sun and moon, it is classed among astronomical problems, of which it is by far the most difficult and its explanation the least satisfactory. The form of the surface of the ocean in equilibrio, when revolving with the earth round its axis, is an ellipsoid flattened at the poles; but the action of the sun and moon, especially of the moon, disturbs the equilibrium of the ocean. If the moon attracted the centre of gravity of the earth and all its particles with equal and parallel forces, the whole system of the earth and the waters that cover it would yield to these forces with a common motion, and the equilibrium of the seas would remain undisturbed. The difference of the forces, and the inequality of their directions alone, trouble the equilibrium.

It is proved by daily experience, as well as by strict mathematical reasoning, that if a number of waves or oscillations be excited in a fluid by different forces, each pursues its course, and has its effect independently of the rest. Now in the tides there are three distinct kinds of oscillations, depending on different causes, and producing their effects independently of each other, which may therefore be estimated separately.

The oscillations of the first kind, which are very small, are independent of the rotation of the earth; and as they depend upon the motion of the disturbing body in its orbit, they are of long periods. The second kind of oscillations depends upon the rotation of the earth, therefore their period is nearly a day; and the oscillations of the third kind vary with an angle equal to twice the angular rotation of the earth; and consequently happen twice in twenty-four hours. The first afford no particular interest, and are extremely small; but the difference of two consecutive tides depends upon the second. At the time of the solstices, this difference, which ought to be very great, according to Newton's theory, is hardly sensible on our shores. La Place has shown that this discrepancy arises from the depth of the sea, and that if the depth were uniform there would be no difference in the consecutive tides but that which is occasioned by local circumstances; it follows, therefore, that as this difference is extremely small, the sea, considered in a large extent, must be nearly of uniform depth, that is to say, there is a certain mean depth from which the deviation is not great. The mean depth of the Pacific Ocean is supposed to be about four miles, that of the Atlantic only three. From the formulæ which determine the difference of the consecutive tides, it is also proved, that the precession of the equinoxes, and the nutation of the earth's axis, are the same as if the sea formed one solid mass with the earth.

Oscillations of the third kind are the semi-diurnal tides, so remarkable on our coasts; they are occasioned by the combined action of the sun and moon, but as the effect of each is independent of the other, they may be considered separately.

The particles of water under the moon are more attracted than the centre of gravity of the earth, in the inverse ratio of the square of the distances; hence they have a tendency to leave the earth, but are retained by their gravitation, which is diminished by this tendency. On the contrary, the moon attracts the centre of the earth more powerfully than she attracts the particles of water in the hemisphere opposite to her; so that the earth has a tendency to leave the waters, but is retained by gravitation, which is again diminished by this tendency. Thus the waters immediately under the moon are drawn from the earth at the same time that the earth is drawn from those which are diametrically opposite to her; in both instances producing an elevation of the ocean of nearly the same height above the surface of equilibrium; for the diminution of the gravitation of the particles in each position is almost the same, on account of the distance of the moon being great in comparison of the radius of the earth. Were the earth entirely covered by the sea, the water thus attracted by the moon would assume the form of an oblong spheroid, whose greater axis would point towards the moon, since the columns of water under the moon and in the direction diametrically opposite to her are rendered lighter in consequence of the diminution of their gravitation; and in order to preserve the equilibrium, the axes 90° distant would be shortened. The elevation, on account of the smaller space to which it is confined, is twice as great as the depression, because the contents of the spheroid always remain the same. The effects of the sun's attraction are in all respects similar to those of the moon's, though greatly less in degree, on account of his distance; he therefore only modifies the form of this spheroid a little. If the waters were capable of assuming the form of equilibrium instantaneously, that is, the form of the spheroid, its summit would always point to the moon, notwithstanding the earth's rotation; but on account of their resistance the rapid motion produced in them by rotation, prevents them from assuming at every instant the form which the equilibrium of the forces acting upon them requires. Hence, on account of the inertia of the waters, if the tides be considered relatively to the whole earth, and open sea, there is a meridian about 30° eastward of the moon, where it is always high water both in the hemisphere where the moon is and in that which is opposite. On the west side of this circle the tide is flowing, on the east it is ebbing, and on every part of the meridian at 90° distant, it is low water. These tides must necessarily happen twice in a day, since the rotation of the earth brings the same point twice under the meridian of the moon in that time, once under the superior, and once under the inferior, meridian.

In the semidiurnal tides there are two phenomena particularly to be distinguished, one occurring twice in a month, and the other twice in a year.

The first phenomenon is, that the tides are much increased in the syzigies, or at the time of new and full moon. In both cases the sun and moon are in the same meridian, for when the moon is new they are in conjunction, and when she is full, they are in opposition. In each of these positions their action is combined to produce the highest or spring tides under that meridian, and the lowest in those points that are 90° distant. It is observed that the higher the sea rises in full tide, the lower it is in the ebb. The neap tides take place when the moon is in quadrature; they neither rise so high nor sink so low as the spring tides. The spring tides are much increased when the moon is in perigee, because she is then nearest to the earth. It is evident that the spring tides must happen twice in a month, since in that time the moon is once new and once full.

The second phenomenon in the tides is the augmentation, which occurs at the time of the equinoxes, when the sun's declination is zero, which happens twice every year. The greatest tides take place when a new or full moon happens near the equinoxes while the moon is in perigee. The inclination of the moon's orbit on the ecliptic is 5° 8′ 47″·9; hence, in the equinoxes, the action of the moon would be increased if her node were to coincide with her perigee. The equinoctial gales often raise these tides to a great height. Besides these remarkable variations, there are others arising from the declination of the sun and moon, which have a great influence on the ebb and flow of the waters. The moon takes about twenty-nine days and a half to vary through all her declinations, which sometimes extend about $28\frac{3}{4}$ degrees on each side of the equator, while the sun requires about $365\frac{1}{4}$ days to accomplish his motion from tropic to tropic through about $23\frac{1}{2}$ degrees, so that their combined motion causes great irregularities, and, at times, their attractive forces counteract each other's effects to a certain extent; but, on an average, the mean monthly range of the moon's declination is nearly the same as the annual range of the declination of the sun; consequently the highest tides take place within the tropics, and the lowest towards the poles.

Both the height and time of high water are thus perpetually changing; therefore, in solving the problem, it is required to determine the heights to which the tides rise, the times at which they happen, and the daily variations. Theory and observation show, that each partial tide increases as the cube of the apparent diameter or of the parallax of the body which produces it, and that it diminishes as the square of the cosine of the declination of that body.

The periodic motions of the waters of the ocean, on the hypothesis of an ellipsoid of revolution entirely covered by the sea, are very far from according with observation; this arises from the very great irregularities in the surface of the earth, which is but partially covered by the sea, from the variety in the depths of the ocean, the manner in which it is spread out on the earth, the position and inclination of the shores, the currents, and the resistance the waters meet with, causes it is impossible to estimate, but which modify the oscillations of the great mass of the ocean. However, amidst all these irregularities, the ebb and flow of the sea maintain a ratio to the forces producing them sufficient to indicate their nature, and to verify the law of the attraction of the sun and moon on the sea. La Place observes, that the investigation of such relations between cause and effect is no less useful in natural philosophy than the direct solution of problems, either to prove the existence of the causes or to trace the laws of their effects. Like the theory of probabilities, it is a happy supplement to the ignorance and weakness of the human mind. Thus the problem of the tides does not admit of a general solution; it is certainly necessary to analyse the general phenomena which ought to result from the attraction of the sun and moon, but these must be corrected in each particular case by local observations modified by the extent and depth of the sea, and the peculiar circumstances of the place.

Since the disturbing action of the sun and moon can only become sensible in a very great extent of water, it is evident that the Pacific Ocean is one of the principal sources of our tides; but, in consequence of the rotation of the earth, and the inertia of the ocean, high water does not happen till some time after the moon's southing. The tide raised in that world of waters is transmitted to the Atlantic, from which sea it moves in a northerly direction along the coasts of Africa and Europe, arriving later and later at each place. This great wave, however, is modified by the tide raised in the Atlantic, which sometimes combines with that from the Pacific in raising the sea, and sometimes is in opposition to it, so that the tides only rise in proportion to their difference. This vast combined wave, reflected by the shores of the Atlantic, extending nearly from pole to pole, still coming northward, pours through the Irish and British Channels into the North Sea, so that the tides in our ports are modified by those of another hemisphere. Thus the theory of the tides in each port, both as to their height and the times at which they take place, is really a matter of experiment, and can only be perfectly determined by the mean of a very great number of observations, including several revolutions of the moon's nodes.

The height to which the tides rise is much

greater in narrow channels than in the open sea, on account of the obstructions they meet with. The sea is so pent up in the British Channel, that the tides sometimes rise as much as fifty feet at St. Malo, on the coast of France, whereas, on the shores of some of the South Sea islands, they do not exceed one or two feet. The winds have a great influence on the height of the tides, according as they conspire with or oppose them; but the actual effect of the wind in exciting the waves of the ocean extends very little below the surface: even in the most violent storms, the water is probably calm at the depth of ninety or a hundred feet. The tidal wave of the ocean does not reach the Mediterranean nor the Baltic, partly from their position and partly from the narrowness of the Straits of Gibraltar and of the Categat, but it is very perceptible in the Red Sea and in Hudson's Bay. In high latitudes, where the ocean is less directly under the influence of the luminaries, the rise and fall of the sea is inconsiderable, so that, in all probability, there is no tide at the poles, or only a small annual and monthly tide. The ebb and flow of the sea are perceptible in rivers to a very great distance from their estuaries. In the Straits of Pauxis, in the river of the Amazons, more than five hundred miles from the sea, the tides are evident. It requires so many days for the tide to ascend this mighty stream, that the returning tides meet a succession of those which are coming up; so that every possible variety occurs in some part or other of its shores, both as to magnitude and time. It requires a very wide expanse of water to accumulate the impulse of the sun and moon, so as to render their influence sensible; on that account, the tides in the Mediterranean and Black Sea are scarcely perceptible.

These perpetual commotions in the waters are occasioned by forces that bear a very small proportion to terrestrial gravitation: the sun's action in raising the ocean is only $\frac{1}{38448000}$ of gravitation at the earth's surface, and the action of the moon is little more than twice as much; these forces being in the ratio of 1 to 2·35333, when the sun and moon are at their mean distances from the earth. From this ratio, the mass of the moon is found to be only $\frac{1}{75}$ of that of the earth. Had the action of the sun on the ocean been exactly equal to that of the moon, there would have been no neap tides, and the spring tides would have been of twice the height which the action of either the sun or moon would have produced separately; a phenomenon depending upon the interference of the undulations.

A stone plunged into a pool of still water occasions a series of waves to advance along the surface, though the water itself is not carried forward, but only rises into heights and sinks into hollows, each portion of the surface being elevated and depressed in its turn. Another stone of the same size, thrown into the water near the first, will occasion a similar set of undulations. Then, if an equal and similar wave from each stone arrive at the same spot at the same time, so that the elevation of the one exactly coincides with the elevation of the other, their united effect will produce a wave twice the size of either; but if one wave precede the other by exactly half an undulation, the elevation of the one will coincide with the hollow of the other, and the hollow of the one with the elevation of the other, and the waves will so entirely obliterate one another, that the surface of the water will remain smooth and level. Hence, if the length of each wave be represented by 1, they will destroy one another at intervals of $\frac{1}{2}$, $\frac{3}{2}$, $\frac{5}{2}$, &c., and will combine their effects at the intervals 1, 2, 3, &c. It will be found, according to this principle, when still water is disturbed by the fall of two equal stones, that there are certain lines on its surface of a hyperbolic form, where the water is smooth in consequence of the waves obliterating each other; and that the elevation of the water in the adjacent parts corresponds to both the waves united. Now, in the spring and neap tides, arising from the combination of the simple soli-lunar waves, the spring tide is the joint result of the combination when they coincide in time and place; and the neap tide happens when they succeed each other by half an interval, so as to leave only the effect of their difference sensible. It is therefore evident that, if the solar and lunar tides were of the same height, there would be no difference, consequently no neap tides, and the spring tides would be twice as high as either separately. In the port of Batsha, in Tonquin, where the tides arrive by two channels, of lengths corresponding to half an interval, there is neither high nor low water, on account of the interference of the waves.

The initial state of the ocean has no influence on the tides; for, whatever its primitive conditions may have been, they must soon have vanished by the friction and mobility of the fluid. One of the most remarkable circumstances in the theory of the tides is the assurance that, in consequence of the density of the sea being only one fifth of the mean density of the earth, and that the earth itself increases in density toward the centre, the stability of the equilibrium of the ocean never can be subverted by any physical cause whatever. A general inundation, arising from the mere instability of the ocean, is therefore impossible. A variety of circumstances, however, tend to produce partial variations in the equilibrium of the seas, which is restored by means of currents. Winds, and the periodical melting of the ice at the poles, occasion temporary water-courses; but by far the most important causes are the centrifugal force induced by the velocity of the earth's rotation and variations in the density of the sea.

The centrifugal force may be resolved into two forces—one perpendicular, and another tangent to the earth's surface. The tangential force, though small, is sufficient to make the fluid particles within the polar circles tend towards the equator, and the tendency is much increased by the immense evaporation in the equatorial regions, from the heat of the sun, which disturbs the equilibrium of the ocean; to this may also be added the superior density of the waters near the poles, partly from their low temperature, and partly from their gravitation being less diminished by the action of the sun and moon than that of the seas of lower latitudes. In consequence of the combination of all these circumstances, two great currents perpetually set from each pole towards the equator; but as they come from latitudes where the rotatory motion of the surface of the earth is very much less than it is between the tropics, on account of their inertia, they do not immediately acquire the velocity with which the solid part of the earth's surface is revolving at the equatorial regions, from whence it follows that, within twenty-five or thirty degrees on each side of the line, the ocean appears to have a general motion from east to west, which is much increased by the action of the trade-winds. This mighty mass of rushing waters, at about the tenth degree of south latitude, is turned towards the north-west by the coast of America, runs through the Gulf of Mexico, and, passing the Straits of Florida at the rate of five miles an hour, forms the well-known current of the Gulf-stream, which sweeps along the whole coast of America, and runs northward as far as the bank of Newfoundland, whence, bending to the east, it flows past the Azores and Canary Islands, till it joins the great westerly current of the tropics about latitude 21° north. According to Humboldt, this great circuit of 3800 leagues, which the waters of the Atlantic are perpetually describing between the parallels of eleven and forty-three degrees of latitude, may be accomplished by any one particle in two years and ten months. Besides this, there are branches of the Gulf-stream, which convey the fruits, seeds, and a portion of the warmth of the tropical climates, to our northern shores.

The general westward motion of the South Sea, together with the south polar current, produce various water-courses in the Pacific and Indian Oceans, according as the one or the other prevails. The western set of the Pacific causes currents to pass on each side of Australia, while the polar stream rushes along the Bay of Bengal; but the westerly current again becomes most powerful towards Ceylon and the Maldives, from whence it stretches by the extremity of the Indian peninsula, past Madagascar, to the most northern point of the continent of Africa, where it mingles with the general motion of the seas. Icebergs are sometimes drifted as far as the Azores from the north pole, and from the south pole they have come even to the Cape of Good Hope. In consequence of the polar current, Sir Edward Parry was obliged to give up his attempt to reach the north pole in the year 1827, because he found that the fields of ice were drifting to the south faster than his party could travel over them to the north.

SECTION XV.

The oscillations of the atmosphere, and the changes in its temperature, are measured by variations in the heights of the barometer and thermometer, but the actual length of the liquid columns in these instruments not only depends upon the force of gravitation, but upon capillary attraction, or the force of cohesion, which is a reciprocal attraction between the molecules of the liquid and those of the tube containing it.

All bodies consist of an assemblage of material particles held in equilibrio by a mutual affinity or cohesive force which tends to unite them, and also by a repulsive force—probably caloric, the principle of heat—which tends to separate them. The intensity of these forces decreases rapidly, as the distance between the atoms augments, and becomes altogether insensible as soon as that distance has acquired a sensible magnitude. The particles of matter are so small, that nothing is known of their form further than the dissimilarity of their different sides in certain cases, which appears from their reciprocal attractions during crystallization being more or less powerful, according to the sides they present to one another. It is evident that the density of substances will depend upon the ratio which the opposing forces of cohesion and repulsion bear to one another.

When particles of the same kind of matter are at such distances from each other, that the cohesion which retains them is insensible, the repulsive principle remains unbalanced, and the particles have a tendency to fly from one another, as in aëriform fluids. If the particles approach sufficiently near to produce equilibrium between the attractive and repulsive forces, but not near enough to admit of any influence from their form, perfect mobility will exist among them, resulting from the similarity of their attractions, and they will offer great resistance when compressed, properties which characterize fluids, in which the repulsive principle is greater than in the gases. When the distance between the particles is still less, solids are formed in consequence of the preponderating force of cohesion; but the nature of their structure will vary, because, at such small distances, the power of the mutual attraction of the particles will depend upon their form, and will be modified by the sides they present to one another during their aggregation.

All the phenomena of capillary attraction depend upon the cohesion of the particles of matter. If a glass tube of extremely fine bore, such as a small thermometer-tube, be plunged into a cup of water or alcohol, the liquid will immediately rise in the tube above the level of that in the cup, and the surface of the little column thus suspended will be concave. If the same tube be plunged into a cup full of mercury, the liquid will also rise in the tube, but it will never attain the level of that in the cup, and its surface will be convex. The elevation or depression of the same liquid in different tubes of the same matter is in the inverse ratio of their internal diameters, and altogether independent of their thickness. Whence it follows that the molecular action is insensible at sensible distances, and that it is only the thinnest possible film of the interior surface of the tubes that exerts a sensible action on the liquid. So much indeed is this the case, that, when tubes of the same bore are completely wetted with water throughout their whole extent, mercury will rise to the same height in all of them, whatever be their thickness or density, because the minute coating of moisture is sufficient to remove the internal column of mercury beyond the sphere of attraction of the tube, and to supply the place of a tube by its own capillary attraction. The forces which produce the capillary phenomena are the reciprocal attraction of the tube and the liquid, and of the liquid particles to one another; and in order that the capillary column may be in equilibrio, the weight of that part of it which rises above or sinks below the level of the liquid in the cup must balance these forces.

The estimation of the action of the liquid is a difficult part of this problem. La Place, Dr. Young, and other mathematicians, have considered the liquid within the tube to be of uniform density; but Poisson, in one of those masterly productions in which he elucidates the most abstruse subjects, has recently proved that the phenomena of capillary attraction depend upon a rapid decrease in the density of the liquid column throughout an extremely small space at its surface. Every indefinitely thin layer of a liquid is compressed by the liquid above it, and supported by that below; its degree of condensation depends upon the magnitude of the compressing force, and as this force decreases rapidly towards the surface, where it vanishes, the density of the liquid decreases also. M. Poisson has shown that, when this force is omitted, the capillary surface becomes plane, and that the liquid in the tube will neither rise above nor sink below the level of that in the cup; but, in estimating the forces, it is also necessary to include the variation in the density of the capillary surface round the edges, from the attraction of the tube.

The direction of the resulting force determines the curvature of the surface of the capillary column. In order that a liquid may be in equilibrio, the force resulting from all the forces acting upon it must be perpendicular to the surface. Now, it appears that, as glass is more dense than water or alcohol, the resulting force will be inclined towards the interior side of the tube, therefore the surface of the liquid must be more elevated next the sides of the tube than in the centre, in order to be perpendicular to it, so that it will be concave, as in the thermometer. But as glass is less dense than mercury, the resulting force will be inclined from the interior side of the tube, so that the surface of the capillary column must be more depressed next the sides of the tube than in the centre, in order to be perpendicular to it, and is consequently convex, as may be perceived in the mercury of the barometer when rising. The absorption of moisture by sponges, sugar, salt, &c., are familiar examples of capillary attraction; indeed the pores of sugar are so minute, that there seems to be no limit to the ascent of the liquid. The phenomena arising from the force of cohesion are innumerable: the spherical form of rain-drops and shot, the rise of liquids between plane surfaces, the difficulty of detaching a plate of glass from the surface of water, the force with which two plane surfaces adhere when pressed together,—are all effects of cohesion, entirely independent of atmospheric pressure, and are included in the same analytical formulæ, which express all the circumstances accurately, although the law according to which the forces of cohesion and repulsion vary is unknown, except that they only extend to insensible distances.

The difference between the forces of cohesion and repulsion is called molecular force, and, when modified by the electrical state of the particles, is the general cause of chemical affinities, which only take place between particles of different kinds of matter, though not under all circumstances. Two substances may indeed be mixed, but they will not combine to form a third substance different from both, unless their component particles unite in definite proportions. That is to say—one volume of one of the substances will unite with one volume of the other, or with two volumes, or with three, &c., so as to form a new substance, but in any other proportions it will only form a mixture of the two. For example, one volume of hydrogen gas will combine with eight volumes of oxygen, and form water; or it will unite with sixteen volumes of oxygen, and form deutoxide of hydrogen; but added to any other volume of oxygen, it will merely be a mixture of the two gases. This law of definite proportion, established by Dalton of Manchester, being universal, is one of the most important discoveries in physical science, and furnishes unhoped-for information with regard to the minute and secret operations of nature in the ultimate particles of matter, whose relative weights are thus made known. It would appear also that matter is not infinitely divisible, and Dr. Wollaston has shown that, in all probability, the atmospheres of the sun and planets, as well as of the earth, consist of ultimate atoms, no longer divisible, and if so, that our atmosphere will only extend to that point where the terrestrial attraction is balanced by the elasticity of the air.

All substances may be compressed by a sufficient force, and are said to be more or less elastic according to the facility with which they regain their volume when the pressure is removed, a property which depends upon the repulsive force of their particles. But the pressure may be so great as to bring the particles near enough to one another to come within the sphere of their cohesive force, and then an aeriform fluid may become a liquid, and a liquid a solid. Mr. Faraday has reduced some of the gases to a liquid state by very great compression; but, although atmospheric air is capable of a great diminution of volume, it always retains its gaseous properties, which resume their primitive volume the instant the pressure is removed, in consequence of the elasticity occasioned by the mutual repulsion of its particles.

SECTION XVI.

The atmosphere is not homogeneous; it appears from analysis that, of 100 parts, 79 are azotic gas, and 21 oxygen, the great source of combustion and animal heat. Besides these, there are three or four parts of carbonic acid gas in 1000 parts of atmospheric air. These proportions are found to be the same at all heights hitherto attained by man. The air is an elastic fluid, resisting pressure in every direction, and is subject to the power of gravitation: for, as the space in the top of the tube of a barometer is a vacuum, the column of mercury suspended by the pressure of the atmosphere on the surface of the cistern is a measure of its weight; consequently, every variation in the density occasions a corresponding rise or fall in the barometrical column. The pressure of the atmosphere is about fifteen pounds on every square inch, so that the surface of the whole globe sustains a weight of 11449000000 hundreds of millions of pounds. Shell-fish, which have the power of producing a vacuum, adhere to the rocks by a pressure of fifteen pounds upon every square inch of contact.

Since the atmosphere is both elastic and heavy, its density necessarily diminishes in ascending above the surface of the earth, for each stratum of air is compressed only by the weight above it; therefore the upper strata are less dense, because they are less compressed than those below them. Whence it is easy to show, supposing the temperature to be constant, that, if the heights above the earth be taken in increasing arithmetical progression,—that is, if they increase by equal quantities, as by a foot or a mile, the densities of the strata of air, or the heights of the barometer, which are proportional to them, will decrease in geometrical progression. For example, at the level of the sea, if the mean height of the barometer be 29·922 inches, at the height of 18000 feet it will be 14·961 inches, or one-half as great; at the height of 36000 feet it will be one-fourth as great; at 54000 feet it will be one-eighth, and so on, which affords a method of measuring the heights of mountains with considerable accuracy, and would be very simple if the decrease in the density of the air were exactly according to the preceding law; but it is modified by several circumstances, and chiefly by the changes of temperature, because heat dilates the air and cold contracts it, the variation being $\frac{1}{480}$ for every degree of Fahrenheit's thermometer. Experience shows that the heat of the air decreases as the height above the surface of the earth increases; and it appears, from recent investigations, that the mean temperature of space is 58° below the freezing point of Fahrenheit, which would probably be the temperature of the surface of the earth also, were it not for the non-conducting power of the air, whence it is enabled to retain the heat of the sun's rays, which the earth imbibes and radiates in all directions. The decrease in heat is very irregular, but from the mean of many observations, it appears to be about 14° or 15° for every 9843 feet, which is the cause of the severe cold and eternal snows on the summits of the Alpine chains. The expansion of the atmosphere from the heat of the sun occasions diurnal variations in the height of the barometer. Of the various methods of computing heights from barometrical measurements, that of Ivory has the advantage of combining accuracy with the greatest simplicity. The most remarkable result of barometrical measurement was recently obtained by Baron Von Humboldt, showing that about eighteen thousand square leagues of the north-west of Asia, including the Caspian Sea and the Lake of Aral, are more than three hundred and twenty feet below the level of the surface of the ocean in a state of mean equilibrium. This enormous basin is similar to some of those large cavities on the surface of the moon, and is attributed, by Humboldt, to the upheaving of the surrounding mountain-chains of the Himalaya, of Kuen-Lun, of Thian-Chan, to those of Armenia, of Erzerum, and of Caucasus, which, by undermining the country to so great an extent, caused it to settle below the usual level of the sea. The very contemplation of the destruction that would ensue from the bursting of any of those barriers which now shut out the sea is fearful. In consequence of the diminished pressure of the atmosphere, water boils at a lower temperature on the mountain-tops than in the valleys, which induced Fahrenheit to propose this mode of observation as a method of ascertaining their heights; but although an instrument was constructed for that purpose by Archdeacon Wollaston, it does not appear to have been much employed.

The atmosphere, when in equilibrio, is an ellipsoid flattened at the poles from its rotation with the earth: in that state its strata are of uniform density at equal heights above the level of the sea, and it is sensibly of finite extent, whether it consists of particles infinitely divisible or not. On the latter hypothesis, it must really be finite, and even if its particles be infinitely divisible, it is known, by experience, to be of extreme tenuity at very small heights. The barometer rises in proportion to the superincumbent pressure. At the level of the sea, in the latitude of 45°, and at the temperature of melting ice, the mean height of the barometer being 29·922 inches, the density of air is to the density of a similar volume of mercury, as 1 to 10477·9, consequently the height of the atmosphere, supposed to be of uniform density, would be about 4·95 miles; but as the density decreases upwards in geometrical progression, it is considerably higher, probably about fifty miles. The air, even on the mountain-tops, is sufficiently rare to diminish the intensity of sound, to affect respiration, and to occasion a loss of muscular strength. The blood burst from the lips and ears of M. de Humboldt as he ascended the Andes, and he experienced the same difficulty in kindling and maintaining a fire at great heights that Marco Polo, the Venetian, did on the mountains of Central Asia. At the height of thirty-seven miles, the atmosphere is still dense enough to reflect the rays of the sun when eighteen degrees below the horizon; and although at the height of fifty miles, the bursting of the meteor of 1783 was heard on earth like the report of a cannon, it only proves the immensity of the explosion of a mass, half a mile in diameter, which could produce a sound capable of penetrating air three thousand times more rare than that we breathe; but even these heights are extremely small when compared with the radius of the earth.

The action of the sun and moon disturbs the equilibrium of the atmosphere, producing oscillations similar to those in the ocean, which ought to occasion periodic variations in the heights of the barometer. These, however, are so extremely small, that their existence in latitudes far removed from the equator is doubtful. M. Arago has lately been even led to conclude that the barometrical variations corresponding to the phases of the moon are the effects of some special cause, totally different from attraction, of which the nature and mode of action are unknown. La Place seems to think that the flux and reflux distinguishable at Paris may be occasioned by the rise and fall of the ocean, which forms a variable base to so great a portion of the atmosphere.

The attraction of the sun and moon has no sensible effect on the trade winds; the heat of the sun occasions these aërial currents, by rarefying the air at the equator, which causes the cooler and more dense part of the atmosphere to rush along the surface of the earth to the equator, while that which is heated is carried along the higher strata to the poles, forming two counter currents in the direction of the meridian. But the rotatory velocity of the air, corresponding to its geographical position, decreases towards the poles; in approaching the equator, it must therefore revolve more slowly than the corresponding parts of the earth, and the bodies on the surface of the earth must strike against it with the excess of their velocity, and, by its reaction, they will meet with a resistance contrary to their motion of rotation: so that the wind will appear, to a person supposing himself to be at rest, to blow in a contrary direction to the earth's rotation, or from east to west, which is the direction of the trade winds.

The equator does not exactly coincide with the line which separates the trade winds north and south of it; that line of separation depends upon the total difference of heat in the two hemispheres, arising from the unequal length of their summers, the distribution of land and water, and other causes. There are many proofs of the existence of a counter current above the trade winds. On the Peak of Teneriffe, the prevailing winds are from the west. The ashes of the volcano of St. Vincent's, in the year 1812, were carried to windward as far as the island of Barbadoes by the upper current. The captain of a Bristol ship declared that, on that occasion, dust from St. Vincent's fell to the depth of five inches on the deck at the distance of 500 miles to the eastward; and light clouds have frequently been seen moving rapidly from west to east, at a very great height above the trade winds, which were sweeping along the surface of the ocean in a contrary direction.

SECTION XVII.

WITHOUT the atmosphere, death-like silence would prevail through nature, for it, in common with all substances, has a tendency to impart vibrations to those in contact with it, therefore undulations received by the air, whether it be from a sudden impulse, such as an explosion, or the vibrations of a musical chord, are propagated equally in every direction, and produce the sensation of sound upon the auditory nerves. In the small undulations of deep water in a calm, the vibrations of the liquid particles are made in the vertical plane, that is, at right angles to the direction of the transmission of the waves; but the vibrations of the particles of air which produce sound differ, being performed in the same direction in which the waves of sound travel. The propagation of sound may be illustrated by a field of corn agitated by a gust of wind; for however irregular the motion of the corn may seem, on a superficial view, it will be found, if the intensity of the wind be constant, that the waves are all precisely similar and equal, and that all are separated by equal intervals, and move in equal times.

A sudden blast depresses each ear equally and successively in the direction of the wind, but in consequence of the elasticity of the stalks and the force of the impulse, each ear not only rises again as soon as the pressure is removed, but bends back nearly as much in the contrary direction, and then continues to oscillate backwards and forwards in equal times like a pendulum, to a less and less extent, till the resistance of the air puts a stop to the motion. These vibrations are the same for every individual ear of corn; yet as their oscillations do not all commence at the same time, but successively, the ears will have a variety of positions at any one instant. Some of the advancing ears will meet others in their returning vibrations, and as the times of oscillation are equal for all, they will be crowded together at regular intervals; between these, there will occur equal spaces where the ears will be few, in consequence of being bent in opposite directions; and at other equal intervals they will be in their natural upright positions; so that over the whole field there will be a regular series of condensations and rarefactions among the ears of corn, separated by equal intervals where they will be in their natural state of density. In consequence of these changes the field will be marked by an alternation of bright and dark bands. Thus the successive waves which fly over the corn with the speed of the wind are totally distinct from, and entirely independent of, the extent of the oscillations of each individual ear, though both take place in the same direction. The length of a wave is equal to the space between two ears precisely in the same state of motion, or which are moving similarly, and the time of the vibration of each ear is equal to that which elapses between the arrival of two successive waves at the same point. The only difference between the undulations of a corn-field and those of the air which produce sound is, that each ear of corn is set in motion by an external cause, and is uninfluenced by the motion of the rest, whereas in air, which is a compressible and elastic fluid, when one particle begins to oscillate, it communicates its vibrations to the surrounding particles, which transmit them to those adjacent, and so on continually. Hence, from the successive vibrations of the particles of air, the same regular condensations and rarefactions take place as in the field of corn, producing waves throughout the whole mass of air, though each molecule, like each individual ear of corn, never moves far from its state of rest. The small waves of a liquid, and the undulations of the air, like waves in the corn, are evidently not real masses moving in the direction in which they are advancing, but merely outlines, motions, or forms rushing along, and comprehending all the particles of an undulating fluid, which are at once in a vibratory state. Or, in other words, an undulation is merely the continued transmission in one direction of particles bearing a relative position to one another. It is thus that an impulse given to any one point of the atmosphere is successively propagated in all directions, in waves diverging as from the centre of a sphere to greater and greater distances, but with decreasing intensity, in consequence of the increasing number of particles of inert matter which the force has to move; like the waves formed in still water by a falling stone, which are propagated circularly all around the centre of disturbance. These successive spherical waves are only the repercussions of the condensations and motions of the first particles to which the impulse was given.

The intensity of sound depends upon the violence and extent of the initial vibrations of air, but whatever they may be, each undulation, when once formed, can only be transmitted straight forwards, and never returns back again unless when reflected by an opposing obstacle. The vibrations of the aërial molecules are always extremely small, whereas the waves of sound vary from a few inches to several feet. The various kinds of musical instruments, the human voice, and that of animals, the singing of birds, the hum of insects, the roar of the cataract, the whistling of the wind, and the other nameless peculiarities of sound, at once show an infinite variety in the modes of aërial vibrations, and the astonishing acuteness and delicacy of the ear, thus capable of appreciating the minutest differences in the laws of molecular oscillation.

All mere noises are occasioned by irregular impulses communicated to the ear, and if they be short, sudden, and repeated beyond a certain degree of quickness, the ear loses the intervals of silence, and the sound appears continuous, because, like the eye, it retains the perception of excitement for a moment after the impulse has ceased. Or, in other words, the auditory nerves continue their vibrations for an extremely short period after the impulse, before they return to a state of repose. Still such sounds will be mere noise; in order to produce a musical sound, the impulses, and, consequently, the undulations of the air, must be all exactly similar in duration and intensity, and must recur after exactly equal intervals of time. The quality of a musical note depends upon the abrupt ness, and its intensity upon the violence and extent of the original impulse. But the whole theory of harmonics is founded upon the pitch which varies with the rapidity of the vibrations. The grave, or low tones are produced by very slow vibrations, which increase in frequency progressively, as the note becomes more acute. When the vibrations of a musical chord, for example, are less than sixteen in a second, it will not communicate a continued sound to the ear; the vibrations or pulses increase in number with the acuteness of the note till, at last, all sense of pitch is lost. The whole extent of human hearing, from the lowest note of the organ to the highest known cry of insects, as of the cricket, includes about nine octaves. All ears, however, are by no means gifted with so great a range of hearing; many people, though not at all deaf, are quite insensible to the cry of the bat or the cricket, while to others it is painfully shrill. According to recent experiments by M. Savart, the human ear is capable of hearing sounds arising from about 24000 vibrations in a second, and is consequently able to appreciate a sound which only lasts the twenty-four thousandth part of a second. All people do not hear the deep sounds alike; that faculty seems to depend upon the frequency of the vibrations, and not on the intensity or loudness.

But, although there are limits to the vibrations of our auditory nerves, Dr. Wollaston, who has investigated this curious subject with his usual originality, observes, that "as there is nothing in the nature of the atmosphere to prevent the existence of vibrations incomparably more frequent than any of which we are conscious, we may imagine that animals, like the Grylli, whose powers appear to commence nearly where ours terminate, may have the faculty of hearing still sharper sounds which we do not know to exist, and that there may be other insects hearing nothing in common with us, but endowed with a power of exciting, and a sense which perceives vibrations of the same nature indeed as those which constitute our ordinary sounds, but so remote, that the animals who perceive them may be said to possess another sense agreeing with our own solely in the medium by which it is excited."

The velocity of sound is uniform, and is independent of the nature, extent, and intensity of the primitive disturbance. Consequently sounds, of every quality and pitch, travel with equal speed; the smallest difference in their velocity is incompatible either with harmony or melody, for notes of different pitches and intensities, sounded together at a little distance, would arrive at the ear in different times; and a rapid succession of notes would produce confusion and discord. But as the rapidity with which sound is transmitted depends upon the elasticity of the medium through which it has to pass, whatever tends to increase the elasticity of the air must also accelerate the motion of sound; on that account its velocity is greater in warm than in cold weather, supposing the pressure of the atmosphere constant. In dry air, at the freezing temperature, sound travels at the rate of 1089 feet in a second, and at 62° of Fahrenheit, its speed is 1090 feet in the same time, or 765 miles an hour, which is about three-fourths of the diurnal velocity of the earth's equator. Since all the phenomena of sound are simple consequences of the physical properties of the air, they have been predicted and computed rigorously by the laws of mechanics. It was found, however, that the velocity of sound, determined by observation, exceeded what it ought to have been theoretically by 173 feet, or about one-sixth of the whole amount. La Place suggested that this discrepancy might arise from the increased elasticity of the air, in consequence of a development of latent heat during the undulations of sound, and the result of calculation fully confirmed the accuracy of his views. The aërial molecules being suddenly compressed give out their latent heat, and as air is too bad a conductor to carry it rapidly off, it occasions a momentary and local rise of temperature, which increasing the consecutive expansion of the air, causes a still greater development of heat, and as it exceeds that which is absorbed in the next rarefaction, the air becomes yet warmer, which favours the transmission of sound. Analysis gives the true velocity of sound, in terms of the elevation of temperature that a mass of air is capable of communicating to itself, by the disengagement of its own latent heat, when it is suddenly compressed in a given ratio. This change of temperature, however, cannot be obtained directly by experiment; but by inverting the problem, and assuming the velocity of sound as given by experiment, it was computed that the temperature of a mass of air is raised nine-tenths of a degree when the compression is equal to $\frac{1}{116}$ of its volume.

Probably all liquids are elastic, though considerable force is required to compress them. Water suffers a condensation of nearly 0·0000496 for every atmosphere of pressure, and is consequently capable of conveying sound even more rapidly than air, the velocity being 4708 feet in a second. A person under water hears sounds made in air feebly, but those produced in water very distinctly. According to the experiments of M. Colladon, the sound of a bell was conveyed under water through the Lake of Geneva to the distance of about nine miles. He also perceived that the progress of sound through water is greatly impeded by the interposition of any object, such as a projecting wall; consequently sound under water resembles light, in having a distinct shadow. It has much less in air, being transmitted all round buildings, or other obstacles, so as to be heard in every direction, though often with a considerable diminution of intensity, as when a carriage turns the corner of a street.

The velocity of sound, in passing through solids, is in proportion to their hardness, and is much greater than in air or water. A sound which takes some time in travelling through the air, passes almost instantaneously along a wire six hundred feet long, consequently it is heard twice,—first as communicated by the wire, and afterwards through the medium of the air. The facility with which the vibrations of sound are transmitted along the grain of a log of wood is well known, indeed they pass through iron, glass, and some kinds of wood at the rate of 18530 feet in a second. The velocity of sound is obstructed by a variety of circumstances, such as falling snow, fog, rain, or any other cause which disturbs the homogeneity of the medium through which it has to pass. Humboldt says, that it is on account of the greater homogeneity of the atmosphere during the night that sounds are then better heard than during the day, when its density is perpetually changing from partial variations of temperature. His attention was called to this subject by the rushing noise of the great cataracts of the Orinoco, which seemed to be three times as loud during the night as in the day, from the plain surrounding the Mission of the Apures. This he illustrated by the celebrated experiment of Chladni. A tall glass, half full of champagne, cannot be made to ring as long as the effervescence lasts; in order to produce a musical note, the glass, together with the liquid it contains, must vibrate in unison as a system, which it cannot do, in consequence of the fixed air rising through the wine and disturbing its homogeneity, because the vibrations of the gas being much slower than those of the liquid, the velocity of the sound is perpetually interrupted. For the same reason, the transmission of sound as well as light is impeded in passing through an atmosphere of variable density. Sir John Herschel, in his admirable Treatise on Sound, thus explains the phenomenon. "It is obvious," he says, "that sound as well as light must be obstructed, stifled, and dissipated from its original direction by the mixture of air of different temperatures, and consequently elasticities; and thus the same cause which produces that extreme transparency of the air at night, which astronomers alone fully appreciate, renders it also more favourable to sound. There is no doubt, however, that the universal and dead silence, generally prevalent at night, renders our auditory nerves sensible to impressions which would otherwise escape notice. The analogy between sound and light is perfect in this as in so many other respects. In the general light of day the stars disappear. In the continual hum of voices, which is always going on by day, and which reach us from all quarters, and never leave the ear time to attain complete tranquillity, those feeble sounds which catch our attention at night make no impression. The ear, like the eye, requires long and perfect repose to attain its utmost sensibility."

Many instances may be brought in proof of the strength and clearness with which sound passes over the surface of water or ice. Lieutenant Foster was able to carry on a conversation across Port Bowen harbour, when frozen, a distance of a mile and a half.

The intensity of sound depends upon the extent of the excursions of the fluid molecules, on the energy of the transient condensations and dilatations, and on the greater or less number of particles which experience these effects; and we estimate that intensity by the impetus of these fluid molecules on our organs, which is consequently as the square of the velocity, and not by their inertia, which is as the simple velocity; for were the latter the case, there would be no sound, because the inertia of the receding waves of air would destroy the equal and opposite inertia of those advancing, whence it may be concluded, that the intensity of sound diminishes inversely as the square of the distance from its origin. In a tube, however, the force of sound does not decay as in open air, unless, perhaps, by friction against the sides. M. Biot found, from a number of highly-interesting experiments which he made on the pipes of the aqueducts in Paris, that a continued conversation could be carried on, in the lowest possible whisper, through a cylindrical tube about 3120 feet long, the time of transmission through that space being 2·79 seconds. In most cases sound diverges in all directions; but a very elegant experiment of Dr. Young's shows that there are exceptions. When a tuning-fork vibrates, its two branches alternately recede from and approach one another; each communicates its vibrations to the air, and a musical note is the consequence. If the fork be held upright, about a foot from the ear, and turned round its axis while vibrating, at every quarter revolution the sound will scarcely be heard, while at the intermediate points it will be strong and clear. This phenomenon is occasioned by the air rushing between the two branches of the fork when they recede from one another, and being squeezed out when they approach, so that it is in one state of motion in the direction in which the fork vibrates, and in another at right angles to it.

It appears from theory as well as daily experience, that sound is capable of reflection from surfaces, according to the same laws as light. Indeed any one who has observed the reflection of the waves from a wall on the side of a river, or very wide canal, after the passage of a steam-boat, will have a perfect idea of the reflection of sound and of light. As every substance in nature is more or less elastic, it may be agitated according to its own law, by the impulse of a mass of undulating air; but reciprocally, the surface by its reaction will communicate its undulations back again into the air. Such reflections produce echos, and as a series of them may take place between two or more obstacles, each will cause an echo of the original sound, growing fainter and fainter till it dies away; because sound, like light, is weakened by reflection. Should the reflecting surface be concave towards a person, the sound will converge towards him with increased intensity, which will be greater still if the surface be spherical and concentric with him. Undulations of sound diverging from one focus of an elliptical shell converge in the other after reflection; consequently a sound from the one will be heard in the other as if it were close to the ear. The rolling noise of thunder has been attributed to reverberation between different clouds, which may possibly be the case to some degree; but Sir John Herschel is of opinion, that an intensely prolonged peal is probably owing to a combination of sounds, because the velocity of electricity being incomparably greater than that of sound, the thunder may be regarded as originating in every point of a flash of lightning at the same instant. The sound from the nearest point will arrive first, and if the flash run in a direct line from a person, the noise will come later and later from the remote points of its path in a continued roar. Should the direction of the flash be inclined, the succession of sounds will be more rapid and intense, and if the lightning describe a circular curve round a person, the sound will arrive from every point at the same instant with a stunning crash. In like manner, the subterranean noises heard during earthquakes, like distant thunder, may arise from the consecutive arrival at the ear of undulations propagated at the same instant from nearer and more remote points; or, if they originate in the same point, the sound may come by different routes through strata of different densities.

Sounds under water are heard very distinctly in the air immediately above, but the intensity decays with great rapidity as the observer goes farther off, and is altogether inaudible at the distance of two or three hundred yards: so that waves of sound, like those of light, in passing from a dense to a rare medium, are not only refracted but suffer total reflection at very oblique incidences.

The laws of interference extend also to sound. It is clear that two equal and similar musical strings will be in unison if they communicate the same number of vibrations to the air in the same time. But if two such strings be so nearly in unison that one performs a hundred vibrations in a second, and the other a hundred and one in the same period,—during the first few vibrations, the two resulting sounds will combine to form one of double the intensity of either, because the aërial waves will sensibly coincide in time and place, but the one will gradually gain on the other, till, at the fiftieth vibration, it will be half an oscillation in advance; then the waves of air which produce the sound being sensibly equal, but the receding part of the one coinciding with the advancing part of the other, they will destroy one another, and occasion an instant of silence. The sound will be renewed immediately after, and will gradually increase till the hundredth vibration, when the two waves will combine to produce a sound double the intensity of either. These intervals of silence and greatest intensity, called beats, will recur every second, but if the notes differ much from one another, the alternations will resemble a rattle; and if the strings be in perfect unison, there will be no beats, since there will be no interference. Thus by interference is meant the coexistence of two undulations, in which the lengths of the waves are the same; and as the magnitude of an undulation may be diminished by the addition of another transmitted in the same direction, it follows, that one undulation may be absolutely destroyed by another, when waves of the same length are transmitted in the same direction, provided that the maxima of the undulations are equal, and that one follows the other by half the length of a wave.

SECTION XVIII.

When the particles of elastic bodies are suddenly disturbed by an impulse, they return to their natural position by a series of isochronous vibrations, whose rapidity, force, and permanency depend upon the elasticity, the form, and the mode of aggregation which unites the particles of the body. These oscillations are communicated to the air, and on account of its elasticity they excite alternate condensations and dilatations in the strata of the fluid nearest to the vibrating body: from thence they are propagated to a distance. A string or wire stretched between two pins, when drawn aside and suddenly let go, will vibrate till its own rigidity and the resistance of the air reduce it to rest. These oscillations may be rotatory, in every plane, or confined to one plane, according as the motion is communicated. In the piano-forte, where the strings are struck by a hammer at one extremity, the vibrations probably consist of a bulge running to and fro from end to end. The vibrations of sonorous bodies, however, are compound. Suppose a vibrating string to give the lowest C of the piano-forte, which is the fundamental note of the string; if it be lightly touched exactly in the middle, so as to retain that point at rest, each half will then vibrate twice as fast as the whole, but in opposite directions; the ventral or bulging segments will be alternately above and below the natural position of the string, and the resulting note will be the octave above C. When a point at a third of the length of the string is kept at rest, the vibration will be three times as fast as those of the whole string, and will give the twelfth above C. When the point of rest is one-fourth of the whole, the oscillations will be four times as fast as those of the fundamental note, and will give the double octave, and so on. Now, if the whole string vibrate freely, a good ear will not only hear the fundamental note, but will detect all the others sounding along with it, though with less and less intensity as the pitch becomes higher. These acute sounds, being connected with the fundamental note by the laws of harmony, are called its harmonics. It is clear, from what has been stated, that the string thus vibrating freely could not give all these harmonics at once, unless it divided itself spontaneously at its aliquot parts into segments in opposite states of vibration, separated by points actually at rest. In proof of this, pieces of paper placed on the string at the half, third, fourth, and other aliquot points, will remain on it during its vibration, but will instantly fly off from any of the intermediate points. Thus, according to the law of co-existing undulations, the whole string and each of its aliquot parts are in different and independent states of vibration at the same time; and as all the resulting notes are heard simultaneously, not only the air, but the ear also, vibrates in unison with each at the same instant. The points of rest, called the nodal points of the string, are a mere consequence of the law of interferences. For if a rope fastened at one end be moved to and fro at the other extremity, so as to transmit a succession of equal waves along it, they will be successively reflected when they arrive at the other end of the rope by the fixed point, and in returning they will occasionally interfere with the advancing waves; and as these opposite undulations will at certain points destroy one another, the point of the rope in which this happens will remain at rest. Thus a series of nodes and ventral segments will be produced, whose number will depend upon the tension and the frequency of the alternate motions communicated to the moveable end. So, when a string fixed at both ends is put in motion by a sudden blow at any point of it, the primitive impulse divides itself into two pulses running opposite ways, which are each totally reflected at the extremities, and, running back again along the whole length, are again reflected at the other ends; and thus they will continue to run backwards and forwards, crossing one another at each traverse, and occasionally interfering so as to produce nodes; so that the motion of a string fastened at both ends consists of a wave or pulse, continually doubled back on itself by reflection at the fixed extremities.

A blast of air passing over the open end of a tube, as over the reeds in Pan's pipes; over a hole in one side, as in the flute; or through the aperture called a reed, with a flexible tongue, as in the clarinet, puts the internal column of air into longitudinal vibrations by the alternate condensations and rarefactions of its particles; at the same time the column spontaneously divides itself into nodes, between which the air also vibrates longitudinally, but with a rapidity proportional to the number of divisions, giving the fundamental note and all its harmonics. The nodes are produced on the principle of interferences, by the reflection of the longitudinal undulations, at the closed end or ends of the pipe, as in the musical string, only that in the one case the undulations are longitudinal, and in the other transverse. Glass and metallic rods, when struck at one end, or rubbed in the direction of their length with a wet finger, vibrate longitudinally, like a column of air, by the alternate condensation and expansion of their constituent particles, which produces a clear and beautiful musical note of a high pitch, on account of the rapidity with which these substances transmit sound. Rods, surfaces, and in general all undulating bodies, resolve themselves into nodes; but in surfaces, the parts which remain at rest during their vibrations are lines, which are curved or plane according to the substance, its form, and the mode of vibration. If a little fine dry sand be strewed over the surface of a plate of glass or metal, ground smooth at the edges, and if undulations be excited by drawing the bow of a violin across its edge, it will emit a musical sound, and the sand will immediately arrange itself in the nodal lines, where alone it will accumulate and remain at rest, because the segments of the surface on each side will be in different states of vibration, the one being always elevated while the other is depressed, and as these two motions meet in the nodal lines, they neutralize one another. These lines vary in form and position with the part where the bow is drawn across, and the point by which the plate is held being at rest, must necessarily be in a nodal line; the smallest variation in the pitch changes the nodal lines. A sound may thus be detected though inaudible. The motion of the sand shows in what direction the vibrations take place: if they be perpendicular to the surface, the sand will be violently tossed up and down, till it finds the points of rest; if they be tangential, the sand will only creep along the surface to the nodal lines. Sometimes the undulations are oblique, or compounded of both the preceding. The air of a room, when thrown into undulations by the continued sound of an organ-pipe, or any other means, divides itself into masses separated by nodal curves of double curvature, such as spirals, on each side of which the air is in opposite states of vibration.

All solids which ring when struck, as bells,

drinking-glasses, gongs, &c. have their shape momentarily and forcibly changed by the blow, and from their elasticity, or tendency to resume their natural form, a series of undulations take place, owing to the alternate condensations and rarefactions of the particles of solid matter. These have also their harmonic tones, and, consequently, nodes. Indeed generally when a rigid system of any form whatever vibrates either transversely or longitudinally, it divides itself into a certain number of parts, which perform their vibrations without disturbing one another. These parts are at every instant in alternate states of undulation, and as the points or lines where they join partake of both, they remain at rest because the opposing motions destroy one another.

All bodies have a tendency to impart their undulations both to the air and to substances in contact with them. A musical string gives a very feeble sound when vibrating alone, on account of the small quantity of air set in motion; but when attached to a sounding-board, as in the harp and piano-forte, it communicates its undulations to that surface, and from thence to every part of the instrument, so that the whole system vibrates isochronously, and by exposing an extensive undulating surface, which transmits its undulations to a great mass of air, the sound is much reinforced.

It is evident that the sounding-board and the whole instrument are agitated at once by all the superposed vibrations excited by the simultaneous or consecutive notes that are sounded, each having its perfect effect independently of the rest. The air, notwithstanding its rarity, is capable of transmitting its undulations when in contact with a body susceptible of admitting and exciting them. It is thus that sympathetic undulations are excited by a body vibrating near insulated tended strings, capable of following its undulations, either by vibrating entire, or by separating themselves into their harmonic divisions. When a tuning-fork receives a blow, and is made to rest upon a piano-forte, during its vibration every string which, either by its natural length, or by its spontaneous sub-divisions, is capable of executing corresponding vibrations, responds in a sympathetic note. Some one or other of the notes of an organ are generally in unison with one of the panes, or with the whole sash of a window, which consequently resound when these notes are sounded. A peal of thunder has frequently the same effect. The sound of very large organ-pipes is generally inaudible till the air be set in motion by the undulations of some of the superior accords, and then its sound becomes extremely energetic. Recurring vibrations occasionally influence each other's periods.

For example: two adjacent organ-pipes, nearly in unison, may force themselves into concord, and two clocks, whose rates differed considerably when separate, have been known to beat together when fixed to the same wall.

Every one is aware of the reinforcement of sound by the resonance of cavities. When singing or speaking near the aperture of a wide-mouthed vessel, the intensity of some one note in unison with the air in the cavity is often augmented to a great degree. Any vessel will resound if a body vibrating the natural note of the cavity be placed opposite to its orifice, and be large enough to cover it; or, at least, to set a large portion of the adjacent air in motion. For the sound will be alternately reflected by the bottom of the cavity and the undulating body at its mouth. The first impulse of the undulating substance will be reflected by the bottom of the cavity, and then by the undulating body, in time to combine with the second new impulse; this reinforced sound will also be twice reflected in time to conspire with the third new impulse; and as the same process will be repeated on every new impulse, each will combine with all its echos to reinforce the sound prodigiously.

Several attempts have been made to imitate the articulation of the letters of the alphabet. About the year 1779, MM. Kratzenstein, of St. Petersburgh, and Kempelen, of Vienna, constructed instruments which articulated many letters, words, and even sentences; Mr. Willis, of Cambridge, has recently adapted cylindrical tubes to a reed, whose length can be varied at pleasure by sliding joints. Upon drawing out the tube, while a column of air from the bellows of an organ is passing through it, the vowels are pronounced in the order *i*, *e*, *a*, *o*, *u*; on extending the tube, they are repeated, after a certain interval, in the inverted order *u*, *o*, *a*, *e*, *i*; after another interval, they are again obtained in the direct order, and so on. When the pitch of the reed is very high, it is impossible to sound some of the vowels, which is in perfect correspondence with the human voice, female singers being unable to pronounce *u* and *o* in their high notes. From the singular discoveries of M. Savart, on the nature of the human voice, and the investigations of Mr. Willis on the mechanism of the larynx, it may be presumed that ultimately the utterance or pronunciation of modern languages will be conveyed, not only to the eye, but also to the ear, of posterity. Had the ancients possessed the means of transmitting such definite sounds, the civilized world would still have responded in sympathetic notes at the distance of hundreds of ages.

SECTION XIX.

The action of the atmosphere on light is not less interesting than the theory of sound, for in consequence of the refractive power of the air, no distant object is seen in its true position.

All the celestial bodies appear to be more elevated than they really are, because the rays of light, instead of moving through the atmosphere in straight lines, are continually inflected towards the earth. Light passing obliquely out of a rare into a denser medium, as from vacuum into air, or from air into water, is bent or refracted from its course towards a perpendicular to that point of the denser surface where the light enters it. In the same medium, the sine of the angle contained between the incident ray and the perpendicular is in a constant ratio to the sine of the angle contained by the refracted ray and the same perpendicular; but this ratio varies with the refracting medium. The denser the medium the more the ray is bent. The barometer shows that the density of the atmosphere decreases as the height above the earth increases; and direct experiments prove, that the refractive power of the air increases with its density; it follows, therefore, that if the temperature be uniform, the refractive power of the air is greatest at the earth's surface and diminishes upwards.

A ray of light from a celestial object falling obliquely on this variable atmosphere, instead of being refracted at once from its course, is gradually more and more bent during its passage through it, so as to move in a vertical curved line, in the same manner as if the atmosphere consisted of an infinite number of strata of different densities. The object is seen in the direction of a tangent to that part of the curve which meets the eye, consequently the apparent altitude of the heavenly bodies is always greater than their true altitude. Owing to this circumstance, the stars are seen above the horizon after they are set, and the day is lengthened from a part of the sun being visible, though he really is behind the rotundity of the earth. It would be easy to determine the direction of a ray of light through the atmosphere, if the law of the density were known; but as this law is perpetually varying with the temperature, the cause is very complicated. When rays pass perpendicularly from one medium into another, they are not bent; and experience shows, that in the same surface, though the sines of the angles of incidence and refraction retain the same ratio, the refraction increases with the obliquity of incidence. Hence it appears, from what precedes, that the refraction is greatest at the horizon, and at the zenith there is none; but it is proved that at all heights above ten degrees, refraction varies nearly as the tangent of the angular distance of the object from the zenith, and wholly depends upon the heights of the barometer and thermometer; for the quantity of refraction at the same distance from the zenith varies nearly as the height of the barometer, the temperature being constant; and the effect of the variation of temperature is to diminish the quantity of refraction by about its 480th part for every degree in the rise of Fahrenheit's thermometer. Not much reliance can be placed on celestial observations within less than ten or twelve degrees of the horizon, on account of irregular variations in the density of the air near the surface of the earth, which are sometimes the cause of very singular phenomena. The humidity of the air produces no sensible effect on its refractive power.

Bodies, whether luminous or not, are only visible by the rays which proceed from them; and as the rays must pass through strata of different densities in coming to us, it follows that, with the exception of stars in the zenith, no object either in or beyond our atmosphere is seen in its true place; but the deviation is so small in ordinary cases, that it causes no inconvenience, though in astronomical and trigonometrical observations a due allowance must be made for the effects of refraction. Dr. Bradley's tables of refraction were formed by observing the zenith distances of the sun at his greatest declinations, and the zenith distances of the pole-star above and below the pole; the sum of these four quantities is equal to 180°, diminished by the sum of the four refractions; whence the sum of the four refractions was obtained; and from the law of the variation of refraction determined by theory, he assigned the quantity due to each altitude. The mean horizontal refraction is about 35′ 6″, and at the height of forty-five degrees it is 58″·36. The effect of refraction upon the same star above and below the pole was noticed by Alhazen, a Saracen astronomer of Spain, in the ninth century; but its existence was known to Ptolemy in the second, though he was ignorant of its quantity.

The refraction of a terrestrial object is estimated differently from that of a celestial body; it is measured by the angle contained between the tangent to the curvilineal path of the ray, where it meets the eye, and the straight line joining the eye and the object. Near the earth's surface the path of the ray may be supposed to be circular; and the angle of this path between tangents at the two extremities of this arc is called the horizontal angle. The quantity of terrestrial refraction is obtained by measuring contemporaneously the elevation of the top of a mountain above a point in the plain at its base, and the depression of that point below the top of the mountain. The distance between these two stations is the chord of the horizontal angle; and it is easy to prove that double the refraction is equal to the horizontal angle, diminished by the difference between the apparent elevation and the apparent depression. Whence it appears that, in the mean state of the atmosphere, the refraction is about the fourteenth part of the horizontal angle.

Some very singular appearances occur from the accidental expansion or condensation of the strata of the atmosphere contiguous to the surface of the earth, by which distant objects, instead of being elevated, are depressed; and sometimes, being at once both elevated and depressed, they appear double, one of the images being direct, and the other inverted. In consequence of the upper edges of the sun and moon being less refracted than the lower, they often appear to be oval when near the horizon. The looming also, or elevation of coasts, mountains and ships, when viewed across the sea, arises from unusual refraction. A friend of the author's, on the plains of Hindostan, saw the whole upper chain of the Himalaya mountains start into view, from a sudden change in the density of the air, occasioned by a heavy shower after a very long course of dry and hot weather. Single and double images of objects at sea, arising from sudden changes of temperature, which are not so soon communicated to the water on account of its density as to the air, occur more rarely, and are of shorter duration than similar appearances on land. In 1818, Captain Scoresby, whose observations on the phenomena of the polar seas are so valuable, recognised his father's ship by its inverted image in the air, although the vessel itself was below the horizon. He afterwards found that she was seventeen miles beyond the horizon, and thirty miles distant. Two images are sometimes seen suspended in the air over a ship, one direct and the other inverted, with their topmasts or their hulls meeting, according as the inverted image is above or below the direct image. Dr. Wollaston has proved that these appearances are owing to the refraction of the rays through media of different densities, by the very simple experiment of looking along a red hot poker at a distant object. Two images are seen, one direct and another inverted, in consequence of the change induced by the heat in the density of the adjacent air. He produced the same effect by a saline or saccharine solution with water and spirit of wine floating upon it.

Many of the phenomena that have been ascribed to extraordinary refraction seem to be occasioned by a partial or total reflection of the rays of light at the surfaces of strata of different densities. It is well known that when light falls obliquely upon the external surface of a transparent medium, as on a plate of glass, or stratum of air, one portion is reflected and the other transmitted, but when light falls very obliquely upon the internal surface, the whole is reflected and not a ray is transmitted; in all cases the angles made by the incident and reflected rays with a perpendicular to the surface being equal. As the brightness of the reflected image depends on the quantity of light, those arising from total reflection must be by far the most vivid. The delusive appearance of water, so well known to African travellers, and to the Arab of the desert, as the Lake of the Gazelles, is ascribed to the reflection which takes place between strata of air of different densities, owing to radiation of heat from the arid sandy plains. The mirage described by Captain Mundy, in his Journal of a Tour in India, probably arises from this cause. 'A deep precipitous valley below us, at the bottom of which I had seen one or two miserable villages in the morning, bore in the evening a complete resemblance to a beautiful lake; the vapour, which played the part of water, ascending nearly half way up the sides of the vale, and on its bright surface trees and rocks being distinctly reflected. I had not been long contemplating the phenomenon, before a sudden storm came on and dropped a curtain of clouds over the scene.'

An occurrence which happened on the 18th of November, 1804, was probably produced by reflection. Dr. Buchan, while watching the rising sun from the cliff about a mile to the east of Brighton, at the instant the solar disc emerged from the surface of the ocean, saw the cliff on which he was standing, a wind-mill, his own figure and that of a friend, depicted immediately opposite to him on the sea. This appearance lasted about ten minutes, till the sun had risen nearly his own diameter above the surface of the waves. The whole then seemed to be elevated into the air and successively vanished. The rays of the sun fell upon the cliff at an incidence of 73° from the perpendicular, and the sea was covered with a dense fog many yards in height, which gradually receded before the rising sun. When extraordinary refraction takes place laterally, the strata of variable density are perpendicular to the horizon, and when it is combined with vertical refraction, the objects are magnified as if seen through a telescope. From this cause, on the 26th of July, 1798, the cliffs of France, fifty miles off, were seen as distinctly from Hastings as if they had been close at hand, and even Dieppe was said to have been visible in the afternoon.

The stratum of air in the horizon is so much thicker and more dense than the stratum in the vertical, that the sun's light is diminished 1300 times in passing through it, which enables us to look at him when setting without being dazzled. The loss of light, and consequently of heat, by the absorbing power of the atmosphere, increases with the obliquity of incidence. Of ten thousand rays falling on its surface, 8123 arrive at a given point of the earth if they fall perpendicularly; 7024 arrive if the angle of direction be fifty degrees; 2831 if it be seven degrees; and only five rays will arrive through a horizontal stratum. Since so great a quantity of light is lost in passing through the atmosphere, many celestial objects may be altogether invisible from the plain, which may be seen from elevated situations. Diminished splendour and the false estimate we make of distance from the number of intervening objects, lead us to suppose the sun and moon to be much larger when in the horizon than at any other altitude, though their apparent diameters are then somewhat less. Instead of the sudden transitions of light and darkness, the reflective power of the air adorns nature with the rosy and golden hues of the Aurora, and twilight. Even when the sun is eighteen degrees below the horizon, a sufficient portion of light remains to show that, at the height of thirty miles, it is still dense enough to reflect light. The atmosphere scatters the sun's rays, and gives all the beautiful tints and cheerfulness of day. It transmits the blue light in greatest abundance; the higher we ascend, the sky assumes a deeper hue, but in the expanse of space, the sun and stars must appear like brilliant specks in profound blackness.

SECTION XX.

It is impossible thus to trace the path of a sunbeam through our atmosphere without feeling a desire to know its nature, by what power it traverses the immensity of space, and the various modifications it undergoes at the surfaces and the interior of terrestrial substances.

Sir Isaac Newton proved the compound nature of white light, as emitted from the sun, by passing a sunbeam through a glass prism, which, separating the rays by refraction, formed a spectrum or oblong image of the sun, consisting of seven colours, red, orange, yellow, green, blue, indigo, and violet; of which the red is the least refrangible, and the violet the most; but when he reunited these seven rays by means of a lens, the compound beam became pure white as before. He insulated each coloured ray, and finding that it was no longer capable of decomposition by refraction, concluded that white light consists of seven kinds of homogeneous light, and that to the same colour the same refrangibility ever belongs, and to the same refrangibility the same colour. Since the discovery of absorbent media, however, it appears that this is not the constitution of the solar spectrum.

We know of no substance that is either perfectly opaque or perfectly transparent; for even gold may be beaten so thin as to be pervious to light; and, on the contrary, the clearest crystal, the purest air or water, stop or absorb its rays when transmitted, and gradually extinguish them as they penetrate to greater depths. On this account, objects cannot be seen at the bottom of very deep water, and many more stars are visible to the naked eye from the tops of mountains than from the valleys. The quantity of light that is incident on any transparent substance is always greater than the sum of the reflected and refracted rays. A small quantity is irregularly reflected in all directions by the imperfections of the polish by which we are enabled to see the surface; but a much greater portion is absorbed by the body. Bodies that reflect all the rays appear white; those that absorb them all seem black; but most substances, after decomposing the white light which falls upon them, reflect some colours and absorb the rest. A violet reflects the violet rays alone, and absorbs the others; scarlet cloth absorbs almost all the colours except red; yellow cloth reflects the yellow rays most abundantly, and blue cloth those that are blue; consequently colour is not a property of matter, but arises from the action of matter upon light. Thus a white ribbon reflects all the rays, but when dyed red, the particles of the silk acquire the property of reflecting the red rays most abundantly and of absorbing the others. Upon this property of unequal absorption, the colours of transparent media depend; for they also receive their colour from their power of stopping or absorbing some of the colours of white light and transmitting others; as, for example, black and red ink, though equally homogeneous, absorb different kinds of rays; and when exposed to the sun, they become heated in different degrees, while pure water seems to transmit all rays equally, and is not sensibly heated by the passing light of the sun. The rich dark light transmitted by a smalt-blue finger-glass is not a homogeneous colour, like the blue or indigo of the spectrum, but is a mixture of all the colours of white light which the glass has not absorbed; and the colours absorbed are such as, mixed with the blue tint, would form white light. When the spectrum of seven colours is viewed through a thin plate of this glass, they are all visible; and when the plate is very thick, every colour is absorbed between the extreme red and the extreme violet, the interval being perfectly black. But if the spectrum be viewed through a certain thickness of the glass intermediate between the two, it will be found that the middle of the red space, the whole of the orange, a great part of the green, a considerable part of the blue, a little of the indigo, and a very little of the violet, vanish, being absorbed by the blue glass; and that the yellow rays occupy a larger space, covering part of that formerly occupied by the orange on one side, and by the green on the other; so that the blue glass absorbs the red light, which, when mixed with the yellow, constitutes orange; and also absorbs the blue light, which when mixed with the yellow forms the part of the green space next to the yellow. Hence, by absorption, green light is decomposed into yellow and blue, and orange light into yellow and red. Consequently the orange and green rays, though incapable of decomposition by refraction, can be resolved by absorption, and actually consist of two different colours possessing the same degree of refrangibility. Difference of colour, therefore, is not a test of difference of refrangibility, and the conclusion deduced by Newton is no longer admissible as a general truth. By this analysis of the spectrum, not only with blue glass but with a variety of coloured media, Sir David Brewster, so justly celebrated for his optical discoveries, has proved, that the solar spectrum consists of three primary colours, red, yellow, and blue, each of which exists throughout its whole extent, but with different degrees of intensity in different parts; and that the superposition of these three produces all the seven hues according as each primary colour is in excess or defect. Since a certain portion of red, yellow, and blue rays constitute white light, the colour of any point of the spectrum may be considered as consisting of the predominating colour at that point mixed with white light; consequently, by absorbing the excess of any colour at any point of the spectrum above what is necessary to form white light, such white light will appear at that point as never mortal eye looked upon before this experiment, since it possesses the remarkable property of remaining the same after any number of refractions, and of being capable of decomposition by absorption alone.

When the prism is very perfect and the sun-beam small, so that the spectrum may be received on a sheet of white paper in its utmost state of purity, it presents the appearance of a riband shaded with all the prismatic colours, having its breadth irregularly striped or subdivided by an indefinite number of dark and sometimes black lines. The greater number of these rayless lines are so extremely narrow that it is impossible to see them in ordinary circumstances. The best method is to receive the spectrum on the object-glass of a telescope, so as to magnify them sufficiently to render them visible. This experiment may also be made, but in an imperfect manner, by viewing a narrow slit between two nearly-closed window-shutters through a very excellent glass prism held close to the eye, with its refracting angle parallel to the line of light. When the spectrum is formed by the sun's rays, either direct or indirect,—as from the sky, clouds, rainbow, moon, or planets,—the black bands are always found to be in the same parts of the spectrum, and under all circumstances to maintain the same relative positions, breadths, and intensities. Similar dark lines are also seen in the light of the stars, in the electric light, and in the flame of combustible substances, though differently arranged, each star and each flame having a system of dark lines peculiar to itself, which remains the same under every circumstance. Dr. Wollaston and Fraunhofer of Munich discovered these lines deficient of rays independently of each other. Fraunhofer found that their number extends to nearly six hundred. From these he selected seven of the most remarkable, and determined their distances so accurately, that they now form standard and invariable points of reference for measuring the refractive powers of different media on the rays of light, which renders this department of optics as exact as any of the physical sciences. The rays that are wanting in the solar spectrum, which occasion the dark lines, are possibly absorbed by the atmosphere of the sun. If they were absorbed by the earth's atmosphere, the very same rays would be wanting in the spectra from the light of the fixed stars, which is not the case, for it has already been stated that the position of the dark lines is not the same in spectra from star-light and from the light of the sun. The solar rays reflected from the moon and planets would most likely be modified also by their atmospheres, but they are not,—for the dark lines have precisely the same positions in the spectra, from the direct and reflected light of the sun.

A perfectly homogeneous colour is very rarely to be found, but the tints of all substances are most brilliant when viewed in light of their own colour. The red of a wafer is much more vivid in red than in white light; whereas, if placed in homogeneous yellow light, it can no longer appear red, because there is not a ray of red in the yellow light; and were it not that the wafer, like all other bodies, whether coloured or not, reflects white light at its outer surface, it would appear absolutely black when placed in yellow light.

After looking steadily for a short time at a coloured object, such as a red wafer, on turning the eyes to a white substance, a green image of the wafer will appear, which is called the accidental colour of red. All tints have their accidental colours:—thus the accidental colour of orange is blue; that of yellow is indigo; of green, reddish-white; of blue, orange-red; of violet, yellow; and of white, black; and *vice versâ*. When the direct and accidental colours are of the same intensity, the accidental is then called the complementary colour, because any two colours are said to be complementary to one another which produce white when combined.

Recent experiments by Plateau of Brussels prove that direct and accidental colours differ essentially. From these it appears that two complementary colours from direct impression, which would produce white when combined, produce black, or extinguish one another by their union, when accidental; and also that the combination of all the tints of the solar spectrum produces white light if they be from a direct impression on the eye, whereas blackness results from a union of the same tints if they be accidental. M. Plateau attributes the phenomena of accidental colours to a reaction of the retina after being excited by direct vision. When the image of an object is impressed on the retina only for a few moments, the picture left is exactly of the same colour with the object, but in an extremely short time the picture is succeeded by the accidental image. If the prevailing impression be a very strong white light, its accidental image is not black, but a variety of colours in succession. With a little attention it will generally be found that, whenever the eye is affected by one prevailing colour, it sees at the same time the accidental colour, in the same manner as in music the ear is sensible at once to the fundamental note and its harmonic sounds. The imagination has a powerful influence on our optical impressions, and has been known to revive the images of highly luminous objects months and even years afterwards.

SECTION XXI.

NEWTON and most of his immediate successors imagined light to be a material substance emitted by all self-luminous bodies in extremely minute particles, moving in straight lines with prodigious velocity, which, by impinging upon the optic nerves, produce the sensation of light. Many of the observed phenomena have been successfully explained by this theory; it seems, however, totally inadequate to account for the following circumstances.

When two equal rays of red light, proceeding from two luminous points, fall upon a sheet of white paper in a dark room, they will produce a red spot on it, which will be twice as bright as either ray would produce singly, provided the difference in the lengths of the two beams, from the luminous points to the red spot on the paper, be exactly the 0·0000258th part of an inch. The same effect will take place if the difference in their lengths be twice, three times, four times, &c., that quantity. But if the difference in the lengths of the two rays be equal to one-half of the 0·0000258th part of an inch, or to its $1\frac{1}{2}$, $2\frac{1}{2}$, $3\frac{1}{2}$, &c. part, the one light will entirely extinguish the other, and will produce absolute darkness on the paper where the united beams fall. If the difference in the lengths of their paths be equal to the $1\frac{1}{4}$, $2\frac{1}{4}$, $3\frac{1}{4}$, &c. of the 0·0000258th part of an inch, the red spot arising from the combined beams will be of the same intensity which one alone would produce. If violet light be employed, the difference in the lengths of the two beams must be equal to the 0·0000157th part of an inch, in order to produce the same phenomena; and for the other colours the difference must be intermediate between the 0·0000258th and the 0·0000157th part of an inch. Similar phenomena may be seen by viewing the flame of a candle through two very fine slits in a card extremely near to one another; or by admitting the sun's light into a dark room through a pin-hole about the fortieth of an inch in diameter, and receiving the image on a sheet of white paper. When a slender wire is held in the light, its shadow consists of a bright white bar or stripe in the middle, with a series of alternate black and brightly coloured stripes on each side. The rays which bend round the wire in two streams are of equal lengths in the middle stripe; it is consequently doubly bright from their combined effect; but the rays which fall on the paper on each side of the bright stripe, being of such unequal lengths as to destroy one another, form black lines. On each side of these black lines the rays are again of such lengths as to combine to form bright stripes, and so on alternately, till the light is too faint to be visible. When any homogeneous light is used, such as red, the alternations are only black and red; but on account of the heterogeneous nature of white light, the black lines alternate with vivid stripes or fringes of prismatic colours, arising from the superposition of systems of alternate black lines and lines of each homogeneous colour. That the alternation of black lines and coloured fringes actually does arise from the mixture of the two streams of light which flow round the wire, is proved by their vanishing the instant one of the streams is interrupted. It may therefore be concluded, as often as these stripes of light and darkness occur, that they are owing to the rays combining at certain intervals to produce a joint effect, and at others to extinguish one another. Now it is contrary to all our ideas of matter to suppose that two particles of it should annihilate one another under any circumstances whatever; while, on the contrary, it is impossible not to be struck with the perfect similarity between the interferences of small undulations of air and water and the preceding phenomena. The analogy is indeed so perfect, that philosophers of the highest authority concur in the supposition that the celestial regions are filled with an extremely rare, imponderable, and highly elastic medium or ether, whose particles are capable of receiving the vibrations communicated to them by self-luminous bodies, and of transmitting them to the optic nerves, so as to produce the sensation of light. The acceleration in the mean motion of Encke's comet renders the existence of such a medium almost certain. It is clear that, in this hypothesis, the alternate stripes of light and darkness are entirely the effect of the interference of the undulations; for, by actual measurement, the length of a wave of the mean red rays of the solar spectrum is equal to the 0·0000258th part of an inch; consequently, when the elevation of the waves combine, they produce double the intensity of light that each would do singly; and when half a wave combines with a whole,—that is, when the hollow of one wave is filled up by the elevation of another, darkness is the result. At intermediate points between these extremes, the intensity of the light corresponds to intermediate differences in the lengths of the rays.

The theory of interferences is a particular case of the general mechanical law of the superposition of small motions; whence it appears that the disturbance of a particle of an elastic medium, produced by two coexistent undulations, is the sum of the disturbances which each undulation would produce separately; consequently the particle will move in the diagonal of a parallelogram, whose sides are the two undulations. If, therefore, the two undulations agree in direction, or nearly so, the resulting motion will be very nearly equal to their sum, and in the same direction: if they nearly oppose one another, the resulting motion will be nearly equal to their difference; and if the undulations be equal and opposite, the resultant will be zero, and the particle will remain at rest.

The preceding experiments, and the inferences deduced from them, which have led to the establishment of the doctrine of the undulations of light, are the most splendid memorials of our illustrious countryman Dr. Thomas Young, though Huygens was the first to originate the idea.

It is supposed that the particles of luminous bodies are in a state of perpetual agitation, and that they possess the property of exciting regular vibrations in the ethereal medium, corresponding to the vibrations of their own molecules; and that, on account of its elastic nature, one particle of the ether, when set in motion, communicates its vibrations to those adjacent, which in succession transmit them to those farther off, so that the primitive impulse is transferred from particle to particle, and the undulating motion darts through ether like a wave in water. Although the progressive motion of light is known by experience to be uniform, and in a straight line, the vibrations of the particles are always at right angles to the direction of the ray. The propagation of light is like the spreading of waves in water; but if one ray alone be considered, its motion may be conceived by supposing a rope of indefinite length stretched horizontally, one end of which is held in the hand. If it be agitated to and fro at regular intervals, with a motion perpendicular to its length, a series of similar and equal tremors or waves will be propagated along it; and if the regular impulses be given in a variety of planes, as up and down, from right to left, and also in oblique directions, the successive undulations will take place in every possible plane. An analogous motion in the ether, when communicated to the optic nerves, would produce the sensation of common light. It is evident that the waves which flow from end to end of the cord in a serpentine form are altogether different from the perpendicular vibratory motion of each particle of the rope, which never deviates far from a state of rest. So in ether each particle vibrates perpendicularly to the direction of the ray; but these vibrations are totally different from, and independent of, the undulations which are transmitted through it, in the same manner as the vibrations of each particular ear of corn are independent of the waves that rush from end to end of a harvest-field when agitated by the wind.

The intensity of light depends upon the amplitude or extent of the vibrations of the particles of ether; while its colour depends upon their frequency. The time of the vibration of a particle of ether is, by theory, as the length of a wave directly, and inversely as its velocity. Now, as the velocity of light is known to be 192000 miles in a second, if the lengths of the waves of the different coloured rays could be measured, the number of vibrations in a second corresponding to each could be computed; but that has been accomplished as follows:—All transparent substances of a certain thickness, with parallel surfaces, reflect and transmit white light, but if they be extremely thin, both the reflected and transmitted light is coloured. The vivid hues on soap-bubbles, the iridescent colours produced by heat on polished steel and copper, the fringes of colour between the laminæ of Iceland spar and sulphate of lime, all consist of a succession of hues disposed in the same order, totally independent of the colour of the substance, and determined solely by its greater or less thickness,—a circumstance which affords the means of ascertaining the length of the waves of each coloured ray, and the frequency of the vibrations of the particles producing them. If a plate of glass be laid upon a lens of almost imperceptible curvature, before an open window, when they are pressed together a black spot will be seen in the point of contact, surrounded by seven rings of vivid colours, all differing from one another. In the first ring, estimated from the black spot, the colours succeed each other in the following order;—black, very faint blue, brilliant white, yellow, orange, and red. They are quite different in the other rings, and in the seventh the only colours are pale bluish-green and very pale pink. That these rings are formed between the two surfaces in apparent contact may be proved by laying a prism on the lens, instead of the plate of glass, and viewing the rings through the inclined side of it that is next to the eye, which arrangement prevents the light reflected from the upper surface mixing with that from the surfaces in contact, so that the intervals between the rings appear perfectly black,—one of the strongest circumstances in favour of the undulatory theory; for, although the phenomena of the rings can be explained by either hypothesis, there is this material difference, that, according to the undulatory theory, the intervals between the rings ought to be absolutely black, which is confirmed by experiment; whereas, by the emanating doctrine, they ought to be half illuminated, which is not found to be the case. M. Fresnel, whose opinion is of the first authority, thought this test conclusive. It may therefore be concluded that the rings arise entirely from the interference of the rays: the light reflected from each of the surfaces in apparent contact reaches the eye by paths of different lengths, and produces coloured and dark rings alternately, according as the reflected waves coincide or destroy one another. The breadths of the rings are unequal; they decrease in width, and the colours become more crowded, as they recede from the centre. Coloured rings are also produced by transmitting light through the same apparatus; but the colours are less vivid, and are complementary to those reflected, consequently the central spot is white.

The size of the rings increases with the obliquity of the incident light; the same colour requiring a greater thickness or space between the glasses to produce it than when the light falls perpendicularly upon them. Now if the apparatus be placed in homogeneous instead of white light, the rings will all be of the same colour with that of the light employed. That is to say, if the light be red, the rings will be red divided by black intervals. The size of the rings varies with the colour of the light. They are largest in red, and decrease in magnitude with the succeeding prismatic colours, being smallest in violet light.

Since one of the glasses is plane and the other spherical, it is evident that, from the point of contact, the space between them gradually increases in thickness all round, so that a certain thickness of air corresponds to each colour, which, in the undulatory system, measures the length of the wave producing it. By actual measurement Sir Isaac Newton found that the squares of the diameters of the brightest parts of each ring are as the odd numbers, 1, 3, 5, 7, &c.; and that the squares of the diameters of the darkest parts are as the even numbers 0, 2, 4, 6, &c. Consequently the intervals between the glasses at these points are in the same proportion. If, then, the thickness of the air corresponding to any one colour could be found, its thickness for all the others would be known. Now, as Sir Isaac Newton knew the radius of curvature of the lens, and the actual breadth of the rings in parts of an inch, it was easy to compute that the thickness of air at the darkest part of the first ring is the $\frac{1}{89000}$th part of an inch, whence all the others have been deduced. As these intervals determine the lengths of the waves on the undulatory hypothesis, it appears that the length of a wave of the extreme red of the solar spectrum is equal to the 0·0000266th part of an inch; that the length of a wave of the extreme violet is equal to the 0·0000167th part of an inch; and as the time of a vibration of a particle of ether producing any particular colour is directly as the length of a wave of that colour, and inversely as the velocity of light, it follows that the molecules of ether producing the extreme red of the solar spectrum perform 458 millions of millions of vibrations in a second; and that those producing the extreme violet accomplish 727 millions of millions of vibrations in the same time. The lengths of the waves of the intermediate colours and the number of their vibrations being intermediate between these two, white light, which consists of all the colours, is consequently a mixture of waves of all lengths between the limits of the extreme red and violet. The determination of these minute portions of time and space, both of which have a real existence, being the actual results of measurement, do as much honour to the genius of Newton as that of the law of gravitation.

The phenomenon of the coloured rings takes place *in vacuo* as well as in air; which proves that it is the distance between the lenses alone, and not the air, which produces the colours. However, if water or oil be put between them, the rings contract, but no other change ensues, and Newton found that the thickness of different media at which a given tint is seen is in the inverse ratio of their refractive indices, so that the thickness of laminæ may be known by their colour, which could not otherwise be measured; and as the position of the colours in the rings is invariable, they form a fixed standard of comparison, well known as Newton's scale of colours; each tint being estimated according to the ring to which it belongs from the central spot inclusively. Not only the periodical colours which have been described, but the colours seen in thick plates of transparent substances, the variable hues of feathers, of insects' wings, and of striated substances, and the coloured fringes surrounding the shadows of all bodies held in an extremely small beam of light, all depend upon the same principle. Whence it appears, that material substances derive their colours from two different causes—some from the law of interference, such as iridescent metals, peacock's feathers, &c., and others from the unequal absorption of the rays of white light, such as vermilion, ultramarine, blue or green cloth, flowers, and the greater number of coloured bodies.

The ethereal medium pervading space is supposed to penetrate all material substances, occupying the interstices between their molecules; but in the interior of refracting media it exists in a state of less elasticity compared with its density in vacuo; and the more refractive the medium the less the elasticity of the ether within it. Hence the waves of light are transmitted with less velocity in such media as glass and water than in the external ether. As soon as a ray of light reaches the surface of a diaphanous reflecting substance, for example, a plate of glass, it communicates its undulations to the ether next in contact with the surface, which thus becomes a new centre of motion, and two hemispherical waves are propagated from each point of this surface; one of which proceeds forward into the interior of the glass, with a less velocity than the incident wave; and the other is transmitted back into the air with a velocity equal to that with which it came. Thus when refracted, the light moves with a different velocity without and within the glass; when reflected, the ray comes and goes with the same velocity. The particles of ether without the glass which communicate their motions to the particles of the dense and less elastic ether within it, are analogous to small elastic balls striking large ones; for some of the motion will be communicated to the large balls, and the small ones will be reflected. The first would cause the refracted wave, and the last the reflected. Conversely, when the light passes from glass to air, the action is similar to large balls striking small ones. The small balls receive a motion which would cause the refracted ray, and the part of the motion retained by the large ones would occasion the reflected wave; so that when light passes through a plate of glass or of any other medium differing in density from the air, there is a reflection at both surfaces. But this difference exists between the two reflections, that one is caused by a vibration in the same direction with that of the incident ray, and the other by a vibration in the opposite direction.

A single wave of air or ether would not produce the sensation of sound or light. In order to excite vision, the vibrations of the molecules of ether must be regular, periodical, and very often repeated; and as the ear continues to be agitated for a short time after the impulse, by which alone a sound becomes continuous, so also the fibres of the retina, according to M. d'Arcet, continue to vibrate for about the eighth part of a second, after the exciting cause has ceased. Every one must have observed when a strong impression is made by a bright light, that the object remains visible for a short time after shutting the eyes, which is supposed to be in consequence of the continued vibrations of the fibres of the retina. It is quite possible that many vibrations may be excited in the ethereal medium incapable of producing undulations in the fibres of the human retina, which yet have a powerful effect on those of other animals or of insects. Such may receive luminous impressions of which we are totally unconscious, and at the same time they may be insensible to the light and colours which affect our eyes, their perceptions beginning where ours end.

SECTION XXII.

In giving a sketch of the constitution of light, it is impossible to omit the extraordinary property of its polarization, 'the phenomena of which,' Sir John Herschel says, 'are so singular and various, that to one who has only studied the common branches of physical optics, it is like entering into a new world, so splendid as to render it one of the most delightful branches of experimental inquiry, and so fertile in the views it lays open of the constitution of natural bodies, and the minuter mechanism of the universe, as to place it in the very first rank of the physico-mathematical sciences, which it maintains by the rigorous application of geometrical reasoning its nature admits and requires.'

In general, when a ray of light is reflected from a pane of plate-glass, or any other substance, it may be reflected a second time from another surface, and it will also pass freely through transparent bodies; but if a ray of light be reflected from a pane of plate-glass at an angle of 57°, it is rendered totally incapable of reflection at the surface of another pane of glass in certain definite positions, but will be completely reflected by the second pane in other positions. It likewise loses the property of penetrating transparent bodies in particular positions, whilst it is freely transmitted by them in others. Light so modified, as to be incapable of reflection and transmission in certain directions, is said to be polarized. This name was originally adopted from an imaginary analogy in the arrangement of the particles of light on the Corpuscular doctrine to the poles of a magnet, and is still retained in the undulatory theory.

Light may be polarized by reflection from any polished surface, and the same property is also imparted by refraction. It is proposed to explain these methods of polarizing light, to give a short account of its most remarkable properties, and to endeavour to describe a few of the splendid phenomena it exhibits.

If a brown tourmaline, which is a mineral generally crystallized in the form of a long prism, be cut longitudinally, that is, parallel to the axis of the prism, into plates about the thirtieth of an inch in thickness, and the surfaces polished, luminous objects may be seen through them, as through plates of coloured glass. The axis of each plate is, in its longitudinal section, parallel to the axes of the prism whence it was cut. If one of these plates be held perpendicularly between the eye and a candle, and turned slowly round in its own plane, no change will take place in the image of the candle; but if the plate be held in a fixed position, with its axis or longitudinal section vertical, when a second plate is interposed between it and the eye, parallel to the first, and turned slowly round in its own plane, a remarkable change will be found to have taken place in the nature of the light, for the image of the candle will vanish and appear alternately at every quarter revolution of the plate, varying through all degrees of brightness down to total, or almost total, evanescence, and then increasing again by the same degrees as it had before decreased. These changes depend upon the relative positions of the plates. When the longitudinal sections of the two plates are parallel, the brightness of the image is at its maximum; and when the axes of the sections cross at right angles, the image of the candle vanishes. Thus the light, in passing through the first plate of tourmaline, has acquired a property totally different from the direct light of the candle. The direct light would have penetrated the second plate equally well in all directions, whereas the refracted ray will only pass through it in particular positions, and is altogether incapable of penetrating it in others. The refracted ray is polarized in its passage through the first tourmaline, and experience shows that it never loses that property, unless when acted upon by a new substance. Thus one of the properties of polarized light is proved to be the incapability of passing through a plate of tourmaline perpendicular to it, in certain positions, and its ready transmission in other positions at right angles to the former.

Many other substances have the property of polarizing light. If a ray of light falls upon a transparent medium which has the same temperature, density and structure throughout every part, as fluids, gases, glass, &c., and a few regularly crystallized minerals, it is refracted into a single pencil of light by the laws of ordinary refraction, according to which the ray, passing through the refracting surface from the object to the eye, never quits a plane perpendicular to that surface. Almost all other bodies, such as the greater number of crystallized minerals, animal and vegetable substances, gums, resins, jellies, and all solid bodies having unequal tensions, whether from unequal temperature or pressure, possess the property of doubling the image or appearance of an object seen through them in certain directions; because a ray of natural light falling upon them is refracted into two pencils which move with different velocities, and are more or less separated, according to the nature of the body and the direction of the incident ray. Iceland spar, a carbonate of lime, which, by its natural cleavage, may be split into the form of a rhombohedron, possesses this property in an eminent degree, as may be seen by passing a piece of paper, with a large pin hole in it, on the side of the spar farthest from the eye. The hole will appear double when held to the light. One of these pencils is refracted according to the same law, as in glass or water, never quitting the plane perpendicular to the refracting surface, and therefore called the ordinary ray; but the other does quit that plane, being refracted according to a different and much more complicated law, and on that account is called the extraordinary ray. For the same reason one image is called the ordinary, and the other the extraordinary image. When the spar is turned round in the same plane, the extraordinary image of the hole revolves about the ordinary image which remains fixed, both being equally bright. But if the spar be kept in one position, and viewed through a plate of tourmaline, it will be found that, as the tourmaline revolves, the images vary in their relative brightness—one increases in intensity till it arrives at a maximum, at the same time that the other diminishes till it vanishes, and so on alternately at each quarter revolution, proving both rays to be polarized ; for in one position the tourmaline transmits the ordinary ray, and reflects the extraordinary, and after revolving 90°, the extraordinary ray is transmitted, and the ordinary ray is reflected. Thus another property of polarized light is, that it cannot be divided into two equal pencils by double refraction, in positions of the doubly refracting bodies, in which a ray of common light would be so divided.

Were tourmaline like other doubly refracting bodies, each of the transmitted rays would be double, but that mineral, when of a certain thickness, after separating the light into two polarized pencils, absorbs one of them, and consequently shows only one image of an object.

The pencils of light, on leaving a doubly refracting substance, are parallel; and it is clear, from the preceding experiments, that they are polarized in planes at right angles to each other. But that will be better understood by considering the change produced in common light by the action of the polarizing body. It has been shown that the undulations of ether, which produce the sensation of common light, are performed in every possible plane, at right angles to the direction in which the ray is moving; but the case is very different after the ray has passed through a doubly refracting substance, like Iceland spar. The light then proceeds in two parallel pencils, whose undulations are still, indeed, transverse to the direction of the rays, but they are accomplished in planes at right angles to one another, analogous to two parallel stretched cords, one of which performs its undulations only in a horizontal plane, and the other in a vertical, or upright plane. Thus the polarizing action of Iceland spar, and of all doubly refracting substances, is, to separate a ray of common light whose waves, or undulations, are in every plane, into two parallel rays, whose waves or undulations lie in planes at right angles to each other. The ray of common light may be assimilated to a round rod, whereas the two polarized rays are like two parallel long flat rulers, one of which is laid horizontally on its broad surface, and the other horizontally on its edge. The alternate transmission and obstruction of one of these flattened beams by the tourmaline is similar to the facility with which a thin sheet of paper, or a card, may be passed between the bars of a grating, or wires of a cage, if presented edgeways, and the impossibility of its passing in a direction transverse to the openings of the bars or wires.

Although it generally happens that a ray of light, in passing through Iceland spar, is separated into two polarized rays; yet there is one direction along which it is refracted in one ray only, and that according to the ordinary law. This direction is called the optic axis. Many crystals and other substances have two optic axes, inclined to each other, along which a ray of light is transmitted in one pencil by the law of ordinary refraction. The extraordinary ray is sometimes refracted towards the optic axis, as in quartz, zircon, ice, &c., which are, therefore, said to be positive crystals; but when it is bent from the optic axis, as in Iceland spar, tourmaline, emerald, beryl, &c., the crystals are negative, which is the most numerous class. The ordinary ray moves with uniform velocity within a doubly refracting substance, but the velocity of the extraordinary ray varies with the position of the ray relatively to the optic axis, being a maximum when its motion within the crystal is at right angles to the optic axis, and a minimum when parallel to it. Between these extremes its velocity varies according to a determinate law.

It had been inferred from the action of Iceland spar on light, that, in all doubly refracting substances, one only of the two rays is turned aside from the plane of ordinary refraction, while the other follows the ordinary law; and the great difficulty of observing the phenomena tended to confirm that opinion. M. Fresnel, however, proved, by a most profound mathematical inquiry, *à priori*, that the extraordinary ray must be wanting in glass and other uncrystallized substances, and that it must necessarily exist in carbonate of lime, quartz, and other bodies having one optic axis, but that, in the numerous class of substances which possess two optic axes, both rays must undergo extraordinary refraction, and consequently that both must deviate from their original plane; and these results have been perfectly confirmed by subsequent experiments. This theory of refraction, which, for generalization, is perhaps only inferior to the law of gravitation, has enrolled the name of Fresnel among those which pass not away, and make his early loss a subject of deep regret to all who take an interest in the higher paths of scientific research.

Panes of glass, if sufficiently numerous, will give a polarized beam by refraction. It appears that, when a beam of common light is partly reflected at, and partly transmitted through, a transparent surface, the reflected and refracted pencils contain equal quantities of polarized light, and that their planes of polarization are at right angles to one another; hence, a pile of panes of glass will give a polarized beam by refraction. For if a ray of common light pass through them, part of it will be polarized by the first plate, the second plate will polarize a part of what passes through it, and the rest will do the same in succession, till the whole beam is polarized, except what is lost by reflection at the different surfaces, or by absorption. This beam is polarized in a plane at right angles to the plane of reflection, that is, at right angles to the plane passing through the incident and reflected ray. But by far the most convenient way of polarizing light is by reflection.

A pane of plate glass laid upon a piece of black cloth, on a table at an open window, will appear of a uniform brightness from the reflection of the sky or clouds; but if it be viewed through a plate of tourmaline, having its axis vertical, instead of being illuminated as before, it will be obscured by a large cloudy spot, having its centre quite dark, which will readily be found by elevating or depressing the eye, and will only be visible when the angle of incidence is 57°, that is, when a line from the eye to the centre of the black spot makes an angle of 33° with the surface of the reflector. When the tourmaline is turned round in its own plane, the dark cloud will diminish, and entirely vanish when the axis of the tourmaline is horizontal, and then every part of the surface of the glass will be equally illuminated. As the tourmaline revolves, the cloudy spot will appear and vanish alternately at every quarter revolution. Thus, when a ray of light is incident on a pane of plate glass at an angle of 57°, the reflected ray is rendered incapable of penetrating a plate of tourmaline whose axis is in the plane of incidence; consequently it has acquired the same character as if it had been polarized by transmission through a plate of tourmaline with its axis at right angles to the plane of reflection. It is found by experience that this polarized ray is incapable of a second reflection at certain angles and in certain positions of the incident plane. For if another pane of plate glass, having one surface blackened, be so placed as to make an angle of 33° with the reflected ray, the image of the first pane will be reflected in its surface, and will be alternately illuminated and obscured at every quarter revolution of the blackened pane, according as the plane of reflection is parallel or perpendicular to the plane of polarization. Since this happens by whatever means the light has been polarized, it evinces another general property of polarized light, which is, that it is incapable of reflection in a plane at right angles to the plane of polarization.

All reflecting surfaces are capable of polarizing light, but the angle of incidence at which it is completely polarized, is different in each substance. It appears that the angle for plate-glass is 57°; in crown-glass it is 56° 55′, and no ray will be completely polarized by water, unless the angle of incidence be 53° 11′. The angles at which different substances polarize light are determined by a very simple and elegant law, discovered by Sir David Brewster, 'That the tangent of the polarizing angle for any medium is equal to the sine of the angle of incidence divided by the sine of the angle of refraction of that medium.' Whence also the refractive power even of an opaque body is known when its polarizing angle has been determined.

Metallic substances, and such as are of high refractive powers, like the diamond, polarize imperfectly.

If a ray polarized by refraction or by reflection from any substance not metallic be viewed through a piece of Iceland spar, each image will alternately vanish and re-appear at every quarter revolution of the spar, whether it revolves from right to left, or from left to right; which shows that the properties of the polarized ray are symmetrical on each side of the plane of polarization.

Although there be only one angle in each substance at which light is completely polarized by one reflection, yet it may be polarized at any angle of incidence by a sufficient number of reflections. For if a ray falls upon the upper surface of a pile of glass at an angle greater or less than the polarizing angle, a part only of the reflected ray will be polarized, but a part of what is transmitted will be polarized by reflection at the surface of the second plate, part at the third, and so on till the whole is polarized. This is the best apparatus; but a plate of glass having its inferior surface blackened, or even a polished table, will answer the purpose.

SECTION XXIII.

Such is the nature of polarized light and the laws it follows; but it is hardly possible to convey an idea of the splendour of the phenomena it exhibits under circumstances which an attempt will now be made to describe.

If light polarized by reflection from a pane of glass be viewed through a plate of tourmaline with its longitudinal section vertical, an obscure cloud, with its centre totally dark, will be seen on the glass. Now let a plate of mica, uniformly about the thirtieth of an inch in thickness, be interposed between the tourmaline and the glass; the dark spot will instantly vanish, and instead of it, a succession of the most gorgeous colours will appear, varying with every inclination of the mica, from the richest reds, to the most vivid greens, blues, and purples. That they may be seen in perfection, the mica must revolve at right angles to its own plane. When the mica is turned round in a plane perpendicular to the polarized ray, it will be found that there are two lines in it where the colours entirely vanish : these are the optic axes of the mica ; which is a doubly refracting substance, with two optic axes, along which light is refracted in one pencil.

No colours are visible in the mica, whatever its position may be with regard to the polarized light, without the aid of the tourmaline, which separates the transmitted ray into two pencils of coloured light complementary to one another, that is, which taken together would make white light; one of these it absorbs and transmits the other: it is therefore called the analyzing plate. The truth of this will appear more readily if a film of sulphate of lime between the twentieth and sixtieth of an inch thick be used instead of the mica. When the film is of uniform thickness, only one colour will be seen when it is placed between the analyzing plate and the reflecting glass; as, for example, red: but when the tourmaline revolves, the red will vanish by degrees, till the film is colourless, then it will assume a green hue, which will increase and arrive at its maximum when the tourmaline has turned through ninety degrees; after that the green will vanish and the red will re-appear, alternating at each quadrant. Whence it appears that the tourmaline separates the light which has passed through the film into a red and a green pencil, and that in one position it absorbs the green and lets the red pass, and in another it absorbs the red and transmits the green. This is proved by analyzing the ray with Iceland spar instead of tourmaline, for since the spar does not absorb the light, two images of the sulphate of lime will be seen, one red and the other green, and these exchange colours every quarter revolution of the spar, the red becoming green and the green red, and where the images overlap, the colour is white, proving the red and green to be complementary to each other. The tint depends on the thickness of the film. Films of sulphate of lime the 0·00124 and 0·01818 of an inch respectively, give white light in whatever position they may be held, provided they be perpendicular to the polarized ray; but films of intermediate thickness will give all colours. Consequently a wedge of sulphate of lime, varying in thickness between the 0·00124 and the 0·01818 of an inch, will appear to be striped with all colours when polarized light is transmitted through it. A change in the inclination of the film, whether of mica or sulphate of lime, is evidently equivalent to a variation in thickness.

When a plate of mica held as close to the eye as possible, at such an inclination as to transmit the polarized ray along one of its optic axes, is viewed through the tourmaline with its axis vertical, a most splendid appearance is presented. The cloudy spot, which is in the direction of the optic axis, is seen surrounded by a set of vividly coloured rings of an oval form, divided into two unequal parts by a black curved band passing through the cloudy spot about which the rings are formed. The other optic axis of the mica exhibits a similar image.

When the two optic axes of a crystal make a small angle with one another, as in nitre, the two sets of rings touch externally; and if the plate of nitre be turned round in its own plane, the black transverse bands undergo a variety of changes, till at last the whole richly coloured image assumes the form of the figure 8, traversed by a black cross. Substances having one optic axis have but one set of coloured circular rings, with a broad black cross passing through its centre and dividing the rings into four equal parts. When the analyzing plate revolves, this figure recurs at every quarter revolution, but in the intermediate positions it assumes the complementary colours, the black cross becoming white.

It is in vain to attempt to describe the beautiful phenomena exhibited by innumerable bodies, all of which undergo periodic changes in form and colour when the analyzing plate revolves, but not one of them shows a trace of colour without the aid of tourmaline or something equivalent to analyze the light, and as it were to call these beautiful phantoms into existence. Tourmaline has the disadvantage of being itself a coloured substance, but that inconvenience may be avoided by employing a reflecting surface as an analyzing plate. When polarized light is reflected by a plate of glass at the polarizing angle, it will be separated into two coloured pencils, and when the analyzing plate is turned round in its own plane, it will alternately reflect each ray at every quarter revolution, so that all the phenomena that have been described will be seen by reflection on its surface.

Coloured rings are produced by analyzing polarized light transmitted through glass melted and suddenly or unequally cooled, also in thin plates of glass bent with the hand, in jelly indurated or compressed, &c. &c.; in short, all the phenomena of coloured rings may be produced, either permanently or transiently, in a variety of substances, by heat and cold, rapid cooling, compression, dilatation, and induration; and so little apparatus is necessary for performing the experiments, that, as Sir John Herschell observes, a piece of window-glass or a polished table to polarize the light, a sheet of clear ice to produce the rings, and a broken fragment of plate-glass placed near the eye to analyze the light, are alone requisite to produce one of the most splendid of optical exhibitions.

It has been observed that when a ray of light, polarized by reflection from any surface not metallic, is analyzed by a doubly refracting substance, it exhibits properties which are symmetrical both to the right and left of the plane of reflection, and the ray is then said to be polarized according to that plane. This symmetry is not destroyed when the ray, before being analyzed, traverses the optic axis of a crystal having but one optic axis, as evidently appears from the circular form of the coloured rings already described. Regularly crystallized quartz, or rock crystal, however, forms an exception. In it, even though the rays should pass through the optic axis itself, where there is no double refraction, the primitive symmetry of the ray is destroyed, and the plane of primitive polarization deviates either to the right or left of the observer, by an angle proportional to the thickness of the plate of quartz. This angular motion, or true rotation of the plane of polarization, which is called circular polarization, is clearly proved by the phenomena. The coloured rings produced by all crystals having but one optic axis are circular, and traversed by a black cross concentric with the rings; so that the light entirely vanishes throughout the space enclosed by the interior ring, because there is neither double refraction nor polarization along the optic axis; but in the system of rings produced by a plate of quartz, whose surfaces are perpendicular to the axis of the crystal, the part within the interior ring, instead of being void of light, is occupied by a uniform tint of red, green, or blue, according to the thickness of the plate. Suppose the plate of quartz to be $\frac{1}{25}$ of an inch thick, which will give the red tint to the space within the interior ring; when the analyzing plate is turned in its own plane through an angle of $17\frac{1}{2}°$, the red hue vanishes. If a plate of rock crystal, $\frac{2}{25}$ of an inch thick, be used, the analyzing plate must revolve through 35° before the red tint vanishes, and so on; every additional 25th of an inch in thickness requiring an additional rotation of $17\frac{1}{2}°$, whence it is manifest that the plane of polarization revolves in the direction of a spiral within the rock crystal. It is remarkable that, in some crystals of quartz, the plane of polarization revolves from right to left, and in others from left to right, although the crystals themselves differ apparently only by a very slight, almost imperceptible, variety in form. In these phenomena, the rotation to the right is accomplished according to the same laws, and with the same energy, as that to the left. But if two plates of quartz be interposed which possess different affections, the second plate undoes, either wholly or partly, the rotatory motion which the first had produced, according as the plates are of equal or unequal thickness. When the plates are of unequal thickness, the deviation is in the direction of the strongest, and exactly the same with that which a third plate would produce equal in thickness to the difference of the two.

M. Biot has discovered the same properties in a variety of liquids. Oil of turpentine and an essential oil of laurel cause the plane of polarization to turn to the left, whereas the syrup of the sugar-cane and a solution of natural camphor by alcohol turn it to the right. A compensation is effected by the superposition or mixture of two liquids which possess these opposite properties, provided no chemical action takes place. A remarkable difference was also observed by M. Biot between the action of the particles of the same substances when in a liquid or solid state. The syrup of grapes, for example, turns the plane of polarization to the left as long as it remains liquid, but as soon as it acquires the solid form of sugar, it causes the plane of polarization to revolve towards the right, a property which it retains even when again dissolved. Instances occur also in which these circumstances are reversed.

A ray of light passing through a liquid possessing the power of circular polarization is not affected by mixing other fluids with the liquid,—such as water, ether, alcohol, &c., which do not possess circular polarization themselves, the angle of deviation remaining exactly the same as before the mixture; whence M. Biot infers that the action exercised by the liquids in question does not depend upon their mass, but that it is a molecular action, exercised by the ultimate particles of matter, which only depends upon their individual constitution, and is entirely independent of the positions and mutual distances of the particles with regard to each other. This peculiar action of matter on light affords the means of detecting varieties in the nature of substances which have eluded chemical research. For example, no chemical difference has been discovered between syrup from the sugar-cane and syrup from grapes; yet the first causes the plane of polarization to revolve to the right, and the other to the left, therefore some essential difference must exist in the nature of their ultimate molecules. The same difference is to be traced between the juices of such plants as give sugar similar to that from the cane and those which give sugar like that obtained from grapes. M. Biot has shown, by these important discoveries, that circular polarization surpasses the power of chemical analysis in giving certain and direct evidence of the similarity or difference existing in the molecular constitution of bodies, as well as of the permanency of that constitution, or of the fluctuations to which it may be liable. This eminent philosopher is now engaged in a series of experiments on the progressive changes in the sap of vegetables at different distances from their roots, and on the products that are formed at the various epochs of vegetation, from their action on polarized light.

One of the many brilliant discoveries of M. Fresnel is the production of circular and elliptical polarization by the internal reflection of light from plate-glass. He has shown that if light, polarized by any of the usual methods, be twice reflected within a glass rhomb of a given form, the vibrations of the ether that are perpendicular to the plane of incidence will be retarded a quarter of a vibration, which causes the vibrating particles to describe a circular helix, or curve, like a corkscrew. However, that only happens when the plane of polarization is inclined at an angle of 45° to the plane of incidence. When these two planes form an angle, either greater or less, the vibrating particles move in an elliptical helix, which curve may be represented by twisting a thread in a spiral about an oval rod. These curves will turn to the right or left according to the position of the incident plane.

The motion of the ethereal medium in elliptical and circular polarization may be represented by the analogy of a stretched cord; for if the extremity of such a cord be agitated at equal and regular intervals by a vibratory motion entirely confined to one plane, the cord will be thrown into an undulating curve lying wholly in that plane. If to this motion there be superadded another, similar and equal, but perpendicular to the first, the cord will assume the form of an elliptical helix; its extremity will describe an ellipse, and every molecule throughout its length will successively do the same. But if the second system of vibrations commence exactly a quarter of an undulation later than the first, the cord will take the form of a circular helix, or corkscrew; the extremity of it will move uniformly in a circle, and every molecule throughout the cord will do the same in succession. It appears, therefore, that both circular and elliptical polarization may be produced by the composition of the motions of two rays in which the particles of ether vibrate in planes at right angles to one another.

Professor Airy, in a very profound and able paper lately published in the Cambridge Transactions, has proved that all the different kinds of polarized light are obtained from rock crystal. When polarized light is transmitted through the axis of a crystal of quartz in the emergent ray, the particles of ether move in a circular helix; and when it is transmitted obliquely, so as to form an angle with the axis of the prism, the particles of ether move in an elliptical helix, the ellipticity increasing with the obliquity of the incident ray; so that, when the incident ray falls perpendicularly to the axis, the particles of ether move in a straight line. Thus quartz exhibits every variety of elliptical polarization, even including the extreme cases where the excentricity is zero, or equal to the greater axis of the ellipse. In many crystals the two rays are so little separated, that it is only from the nature of the transmitted light that they are known to have the property of double refraction. M. Fresnel discovered, by experiments on the properties of light passing through the axis of quartz, that it consists of two superposed rays moving with different velocities; and Professor Airy has proved that, in these two rays, the molecules of ether vibrate in similar ellipses at right angles to each other, but in different directions; that their ellipticity varies with the angle which the incident ray makes with the axis; and that, by the composition of their motions, they produce all the phenomena of the polarized light observed in quartz.

It appears from what has been said, that the molecules of ether always perform their vibrations at right angles to the direction of the ray, but very differently in the various kinds of light. In natural light the vibrations are rectilinear, and in every plane; in ordinary polarized light they are rectilinear, but confined to one plane; in circular polarization the vibrations are circular; and in elliptical polarization the molecules vibrate in ellipses. These vibrations are communicated from molecule to molecule in straight lines when they are rectilinear, in a circular helix when they are circular, and in an oval or elliptical helix when elliptical.

Some fluids possess the property of circular polarization, as oil of turpentine; and elliptical polarization, or something similar, seems to be produced by reflection from metallic surfaces.

The coloured images from polarized light arise from the interference of the rays. MM. Fresnel and Arago proved by experiment that two rays of polarized light interfere and produce coloured fringes if they be polarized in the same plane, but that they do not interfere when polarized in different planes. In all intermediate positions, fringes of intermediate brightness are produced. The analogy of a stretched cord will show how this happens. Suppose the cord to be moved backwards and forwards horizontally at equal intervals: it will be thrown into an undulating curve lying all in one plane. If to this motion there be superadded another, similar and equal, commencing exactly half an undulation later than the first, it is evident that the direct motion every molecule will assume, in consequence of the first system of waves, will at every instant be exactly neutralized by the retrogade motion it would take in virtue of the second; and the cord itself will be quiescent, in consequence of the interference. But if the second system of waves be in a plane perpendicular to the first, the effect would only be to twist the rope, so that no interference would take place. Rays polarized at right angles to each other may subsequently be brought into the same plane without acquiring the property of producing coloured fringes; but if they belong to a pencil, the whole of which was originally polarized in the same plane, they will interfere.

The manner in which the coloured rays are formed may be conceived by considering that, when polarized light passes through the optic axis of a doubly refracting substance,—as mica, for example,—it is divided into two pencils by the analyzing tourmaline; and as one ray is absorbed, there can be no interference. But when the polarized light passes through the mica in any other direction, it is separated into two white rays, and these are again divided into four pencils by the tourmaline, which absorbs two of them; and the other two, being transmitted in the same plane, with different velocities, interfere and produce the coloured phenomena. If the analysis be made with Iceland spar, the single ray passing through the optic axis of the mica will be refracted into two rays polarized in different planes, and no interference will happen: but when two rays are transmitted by the mica, they will be separated into four by the spar, two of which will interfere to form one image, and the other two, by their interference, will produce the complementary colours of the other image, when the spar has revolved through 90°; because, in such positions of the spar as produce the coloured images, only two rays are visible at a time, the other two being reflected. When the analysis is accomplished by reflection, if two rays are transmitted by the mica, they are polarized in planes at right angles to each other; and if the plane of reflection of either of these rays be at right angles to the plane of polarization, only one of them will be reflected, and therefore no interference can take place; but in all other positions of the analyzing plate, both rays will be reflected in the same plane, and consequently will produce coloured rings by their interference.

It is evident that a great deal of the light we see must be polarized, since most bodies which have the power of reflecting or refracting light also have the power of polarizing it. The blue light of the sky is completely polarized at an angle of 74° from the sun in a plane passing through his centre.

A constellation of talent, almost unrivalled at any period in the history of science, has contributed to the theory of polarization, though the original discovery of that property of light was accidental, and arose from an occurrence which, like thousands of others, would have passed unnoticed, had it not happened to one of those rare minds capable of drawing the most important inferences from circumstances apparently trifling. In 1808, while M. Malus was accidentally viewing, with a doubly refracting prism, a brilliant sunset reflected from the windows of the Luxembourg palais in Paris, on turning the prism slowly round, he was surprised to see a very great difference in the intensity of the two images, the most refracted alternately changing from brightness to obscurity at each quadrant of revolution. A phenomenon so unlooked for induced him to investigate its cause, whence sprung one of the most elegant and refined branches of physical optics.

SECTION XXIV.

THE numerous phenomena of periodical colours arising from the interference of light, which do not admit of satisfactory explanation on any other principle than the undulatory theory, are the strongest arguments in favour of that hypothesis; and even cases which at one time seemed unfavourable to that doctrine have proved, upon investigation, to proceed from it alone. Such is the erroneous objection which has been made in consequence of a difference in the mode of action of light and sound under the same circumstances in one particular instance. When a ray of light from a luminous point, and a diverging sound, are both transmitted through a very small hole into a dark room, the light goes straight forward, and illuminates a small spot on the opposite wall, leaving the rest in darkness; whereas the sound, on entering, diverges in all directions, and is heard in every part of the room. These phenomena, however, instead of being at variance with the undulatory theory, are direct consequences of it, arising from the very great difference between the magnitude of the undulations of sound and those of light. The undulations of light are incomparably less than the minute aperture, while those of sound are much greater; therefore, when light, diverging from a luminous point, enters the hole, the rays round its edges are oblique, and consequently of different lengths, while those in the centre are direct, and nearly or altogether of the same lengths; so that the small undulations between the centre and the edges are in different phases, that is, in different states of undulation; and therefore the greater number of them interfere, and, by destroying one another, produce darkness all around the edges of the aperture; whereas the central rays, having the same phases, combine and produce a spot of bright light on a wall or screen directly opposite the hole. The waves of air producing sound, on the contrary, being very large compared with the hole, do not sensibly diverge in passing through it, and are therefore all so nearly of the same length, and consequently in the same phase, or state of undulation, that none of them interfere sufficiently to destroy one another; hence all the particles of air in the room are set into a state of vibration, so that the intensity of the sound is very nearly everywhere the same. It is probable, however, that, if the aperture were large enough, sound diverging from a point without would scarcely be audible, except immediately opposite the opening. Strong as the preceding cases may be, the following experiment, recently published by Professor Airy, seems to be decisive in favour of the undulatory doctrine. Suppose a plano-convex lens of very great radius to be placed upon a plate of very highly polished metal. When a ray of polarized light falls upon this apparatus at a very great angle of incidence, Newton's rings are seen at the point of contact. But as the polarizing angle of glass differs from that of metal, when the light falls on the lens at the polarizing angle of glass, the black spot and the system of rings vanish: for although light in abundance continues to be reflected from the surface of the metal, not a ray is reflected from the surface of the glass that is in contact with it, consequently no interference can take place; which proves, beyond a doubt, that Newton's rings result from the interference of the light reflected from the surfaces apparently in contact.

Notwithstanding the successful adaptation of the undulatory system to phenomena, it cannot be denied that an objection still exists in the dispersion of light, unless the explanation given by Professor Airy be deemed sufficient. A sunbeam falling on a prism, instead of being refracted to a single point, is dispersed, or scattered over a considerable space, so that the rays of the coloured spectrum, whose waves are of different lengths, have different degrees of refrangibility, and consequently move with different velocities, either in the medium which conveys the light from the sun, or in the refracting medium, or in both; whereas it has been shown that rays of all colours move with the same velocity. If, indeed, the velocities of the various rays were different in space, the aberration of the fixed stars, which is inversely as the velocity, would be different for different colours, and every star would appear as a spectrum whose length would be parallel to the direction of the earth's motion, which is not found to agree with observation. Besides, there is no such difference in the velocities of the long and short waves of air in the analogous case of sound, since notes of the lowest and highest pitch are heard in the order in which they are struck. The solution of this anomalous case suggested by Professor Airy from a similar instance in the theory of sound, already mentioned, will be best understood in his own words. 'We have every reason,' he observes, 'to think that a part of the velocity of sound depends upon the circumstance that the law of elasticity of the air is altered by the instantaneous development of latent heat on compression, or the contrary effect on expansion. Now, if this heat required time for its development, the quantity of heat developed would depend upon the time during which the particles remained in nearly the same relative state, that is, on the time of vibration. Consequently, the law of elasticity would be different for different times of vibration, or for different lengths of waves; and therefore the velocity of transmission would be different for waves of different lengths. If we suppose some cause which is put in action by the vibration of the particles to affect in a similar manner the elasticity of the medium of light, and if we conceive the degree of development of that cause to depend upon time, we shall have a sufficient explanation of the unequal refrangibility of different coloured rays.' Even should this view be objectionable, instead of being surprised that one discrepant case should occur, it is astonishing to find the theory so nearly complete, if it be considered that no subject in the whole course of physico-mathematical inquiry is more abstruse than the doctrine of the propagation of motion through elastic media, perpetually requiring the aid of analogy from the unconquerable difficulties of the subject.

SECTION XXV.

It is not by vision alone that a knowledge of the sun's rays is acquired,—touch proves that they have the power of raising the temperature of substances exposed to their action; and experience likewise teaches that remarkable changes are effected by their chemical agency. Sir William Herschel discovered that rays of caloric, which produce the sensation of heat, exist independently of those of light; when he used a prism of flint glass, he found the warm rays most abundant in the dark space a little beyond the red extremity of the solar spectrum, from whence they decrease towards the violet, beyond which they are insensible. It may therefore be concluded that the calorific rays vary in refrangibility, and that those beyond the extreme red are less refrangible than any rays of light. Wollaston, Ritter, and Beckman discovered simultaneously that invisible rays, known only by their chemical action, exist in the dark space beyond the extreme violet, where there is no sensible heat: these are more refrangible than any of the rays of light or heat, and gradually decrease in refrangibility towards the other end of the spectrum, where they cease. Thus the solar spectrum is proved to consist of five superposed spectra, only three of which are visible—the red, yellow, and blue; each of the five varies in refrangibility and intensity throughout the whole extent, the visible part being overlapped at one extremity by the chemical, and at the other by the calorific rays. The action of the chemical rays blackens the salts of silver, and their influence is daily seen in the fading of vegetable colours: what object they are destined to accomplish in the economy of nature remains unknown, but certain it is, that the very existence of the animal and vegetable creation depends upon the calorific rays. That the heat-producing rays exist independently of light is a matter of constant experience in the abundant emission of them from boiling water, yet there is every reason to believe that both the calorific and chemical rays are modifications of the same agent which produces the sensation of light. The rays of heat are subject to the same laws of reflection and refraction with those of light; they pass through the gases with the same facility, but a remarkable difference obtains in the transmission of light and heat through most solid and liquid substances, the same body being often perfectly transparent to the luminous, and altogether impermeable to the calorific rays. The experiments of M. de Laroche show that glass, however thin, totally intercepts the obscure rays of caloric when they flow from a body whose temperature is lower than that of boiling water; that, as the temperature increases, the calorific rays are transmitted more and more abundantly; and when the body becomes highly luminous, that they penetrate the glass with perfect ease. The very feeble heat of moonlight must be incapable of penetrating glass, consequently it does not sensibly affect the thermometer, even when concentrated; and, on the contrary, the extreme brilliancy of the sun is probably the reason why his heat, when brought to a focus by a lens, is more intense than any that can be produced artificially; and it is owing to the same cause that glass screens, which entirely exclude the heat of a common fire, are permeable by the solar caloric.

The results of de Laroche have been confirmed by the recent experiments of M. Melloni, whence it appears that the calorific rays pass less abundantly, not only through glass, but through rock-crystal, Iceland spar, and other diaphanous bodies, both solid and liquid, according as the temperature of their origin is diminished, and that they are altogether intercepted when the temperature is about that of boiling water. It is singular that transparency with regard to light is totally different from the power of transmitting heat. In bodies possessing the same degree of transparency for light, the quantities of heat which they transmit differ immensely, though proceeding from the same source. The transmissive power of certain substances having a dark colour exceeds by four or five times that of others perfectly diaphanous, and the calorific rays pass instantaneously through black glass perfectly opaque to light.

The property of transmitting the calorific rays diminishes, to a certain degree, with the thickness of the body they have to traverse, but not so much as might be expected: a piece of very transparent alum transmitted three or four times less radiant heat from the flame of a lamp than a piece of nearly opaque quartz about a hundred times as thick. However, the influence of thickness upon the phenomena of transmission increases with the decrease of temperature in the origin of the rays, and becomes very great when that temperature is low—a circumstance intimately connected with the law established by de Laroche, for M. Melloni observed that the differences between the quantities of caloric transmitted by the same plate of glass, exposed successively to several sources of heat, diminished with the thinness of the plate, and vanished altogether at a certain limit, and that a film of mica transmitted the same quantity of caloric whether it was exposed to incandescent platina or to a mass of iron heated to 360°.

Since the power of penetrating glass increases in proportion as the radiating caloric approaches the state of light, it seemed to indicate that the same principle takes the form of light or heat according to the modification it receives, and that the hot rays are only invisible light, and light luminous caloric; and it was natural to infer that, in the gradual approach of invisible caloric to the condition and properties of luminous caloric, the invisible rays must at first be analogous to the least calorific part of the spectrum, which is at the violet extremity, an analogy which appeared to be greater, by all flame being at first violet or blue, and only becoming white when it has attained the greatest intensity. Thus, as diaphanous bodies transmit light with the same facility whether proceeding from the sun or from a glow-worm, and that no substance had hitherto been found which instantaneously transmits radiant caloric coming from a source of low temperature, it was concluded that no such substance exists, and the great difference between the transmission of light and radiant heat was thus referred to the nature of the agent of heat, and not to the action of matter upon the calorific rays. M. Melloni has, however, discovered in rock-salt a substance which transmits radiant heat with the same facility whether it originates in the brightest flame or luke-warm water, and which consequently possesses the same permeability with regard to heat that all diaphanous bodies have for light. It follows, therefore, that the impermeability of glass and other substances for heat arises from their action upon the calorific rays, and not from the principle of heat. But, although this discovery changes the received ideas drawn from de Laroche's experiments, it establishes a new and unlooked-for analogy between these two great agents of nature. The probability of light and heat being modifications of the same principle is not diminished by the calorific rays being unseen, for the condition of visibility or invisibility may only depend upon the construction of our eyes, and not upon the nature of the agent which produces these sensations in us. The sense of seeing, like that of hearing, may be confined within certain limits; the chemical rays beyond the violet end of the spectrum may be too rapid or not sufficiently excursive in their vibrations to be visible to the human eye; and the calorific rays beyond the other end of the spectrum may not be sufficiently rapid or too extensive in their undulations to affect our optic nerves, though both may be visible to certain animals or insects. We are altogether ignorant of the perceptions which direct the carrier-pigeon to his home, and the vulture to his prey, before he himself is visible even as a speck in the heavens; or of those in the antennæ of insects which warn them of the approach of danger: so likewise beings may exist on earth, in the air, or in the waters, which hear sounds our ears are incapable of hearing, and which see rays of light and heat of which we are unconscious. Our perceptions and faculties are limited to a very small portion of that immense chain of existence which extends from the Creator to evanescence. The identity of action under similar circumstances is one of the strongest arguments in favour of the common nature of the chemical, visible, and calorific rays. They are all capable of reflection from polished surfaces, of refraction through diaphanous substances, of polarization by reflection and by doubly refracting crystals; none of these rays add sensibly to the weight of matter; their velocity is prodigious, they may be concentrated and dispersed by convex and concave mirrors; light and heat pass with equal facility through rock-salt, and both are capable of radiation; the chemical rays are subject to the same law of interference with those of light; and although the interference of the calorific rays has not yet been proved, there is no reason to suppose that they differ from the others in this instance. As the action of matter in so many cases is the same on the whole assemblage of rays, visible and invisible, which constitute a solar beam, it is more than probable that the obscure, as well as the luminous part, is propagated by the undulations of an imponderable ether, and consequently comes under the same laws of analysis.

Liquids, the various kinds of glass, and probably all substances, whether solid or liquid, that do not crystallize regularly, are more pervious to the calorific rays according as they possess a greater refracting power. For example, the chlorid of sulphur, which has a high refracting power, transmits more of the calorific rays than the oils which have a less refracting power: oils transmit more radiant heat than the acids, the acids more than aqueous solutions, and the latter more than pure water, which, of all the series, has the least refracting power, and is the least pervious to heat. M. Melloni observed also that each ray of the solar spectrum follows the same law of action with that of terrestrial rays having their origin in sources of different temperatures, so that the very refrangible rays may be compared to the heat emanating from a focus of high temperature, and the least refrangible to the heat which comes from a source of low temperature. Thus, if the calorific rays emerging from a prism be made to pass through a layer of water contained between two plates of glass, it will be found that these rays suffer a loss in passing through the liquid as much greater as their refrangibility is less. The rays of heat that are mixed with the blue or violet light pass in great abundance, while those in the obscure part which follows the red light are almost totally intercepted. The first, therefore, act like the heat of a lamp, and the last like that of boiling water.

These circumstances explain the phenomena observed by several philosophers with regard to the point of greatest heat in the solar spectrum, which varies with the substance of the prism. It has already been observed that Sir William Herschel, who employed a prism of flint glass, found that point to be a little beyond the red extremity of the spectrum, but, according to M. Seebeck, it is found to be upon the yellow, upon the orange, on the red, or at the dark limit of the red, according as the prism consists of water, sulphuric acid, crown or flint glass. If it be recollected that, in the spectrum from crown glass, the maximum heat is in the red part, and that the solar rays, in traversing a mass of water, suffer losses inversely as their refrangibility, it will be easy to understand the reason of the phenomenon in question. The solar heat which comes to the anterior face of the prism of water consists of rays of all degrees of refrangibility. Now, the rays possessing the same index of refraction with the red light suffer a greater loss in passing through the prism than the rays possessing the refrangibility of the orange light, and the latter lose less in their passage than the heat of the yellow. Thus, the losses, being inversely proportional to the degree of refrangibility of each ray, cause the point of maximum heat to tend from the red towards the violet, and therefore it rests upon the yellow part. The prism of sulphuric acid, acting similarly, but with less energy than that of water, throws the point of greatest heat on the orange; for the same reason the crown and flint glass prisms transfer that point respectively to the red and to its limit. M. Melloni, observing that the maximum point of heat is transferred farther and farther towards the red end of the spectrum, according as the substance of the prism is more and more permeable to heat, inferred that a prism of rock-salt, which possesses a greater power of transmitting the calorific rays than any known body, ought to throw the point of greatest heat to a considerable distance beyond the visible part of the spectrum—an anticipation which experiment fully confirmed, by placing it as much beyond the dark limit of the red rays as the red part is distant from the bluish-green band of the spectrum.

When radiant heat falls upon a surface, part of it is reflected and part of it is absorbed, consequently the best reflectors possess the least absorbing powers. The absorption of the sun's rays is the cause both of the colour and temperature of solid bodies. A black substance absorbs all the rays of light, and reflects none; and since it absorbs at the same time all the calorific rays, it becomes sooner warm, and rises to a higher temperature, than bodies of any other colour. Blue bodies come next to black in their power of absorption. Of all the colours of the solar spectrum, the blue possesses least of the heating power; and since substances of a blue tint absorb all the other colours of the spectrum, they absorb by far the greatest part of the calorific rays, and reflect the blue where they are least abundant. Next in order come the green, yellow, red, and, last of all, white bodies, which reflect nearly all the rays both of light and heat. The temperature of very transparent fluids is not raised by the passage of the sun's rays, because they do not absorb any of them, and as his heat is very intense, transparent solids arrest a very small portion of it.

Rays of heat proceed in diverging straight lines from each point in the surfaces of hot bodies, in the same manner as diverging rays of light dart from every point of the surfaces of those that are luminous. Heated substances, when exposed to the open air, continue to radiate caloric till they become nearly of the temperature of the surrounding medium. The radiation is very rapid at first, but diminishes, according to a known law, with the temperature of the heated body. It appears also that the radiating power of a surface is inversely as its reflecting power; and bodies that are most impermeable to heat radiate least. According to the experiments of Sir John Leslie, radiation proceeds not only from the surfaces of substances, but also from the particles at a minute depth below it. He found that the emission is most abundant in a direction perpendicular to the radiating surface, and is more rapid from a rough than from a polished surface: radiation, however, can only take place in air and in vacuo; it is altogether insensible when the hot body is inclosed in a solid or liquid. All substances may be considered to radiate caloric, whatever their temperature may be, though with different intensities, according to their nature, the state of their surfaces, and the temperature of the medium into which they are brought. But every surface absorbs, as well as radiates, caloric; and the power of absorption is always equal to that of radiation, for it is found that, under the same circumstances, matter which becomes soon warm also cools rapidly. There is a constant tendency to an equal diffusion of caloric, since every body in nature is giving and receiving it at the same instant; each will be of uniform temperature when the quantities of caloric given and received during the same time are equal, that is, when a perfect compensation takes place between each and all the rest. Our sensations only measure comparative degrees of heat: when a body, such as ice, appears cold, it imparts fewer calorific rays than it receives; and when a substance seems to be warm,—for example, a fire,—it gives more caloric than it takes. The phenomena of dew and hoar-frost are owing to this inequality of exchange, for the caloric radiated during the night by substances on the surface of the earth into a clear expanse of sky is lost, and no return is made from the blue vault, so that their temperature sinks below that of the air, from whence they abstract a part of that caloric which holds the atmospheric humidity in solution, and a deposition of dew takes place. If the radiation be great, the dew is frozen, and becomes hoar-frost, which is the ice of dew. Cloudy weather is unfavourable to the formation of dew, by preventing the free radiation of caloric, and actual contact is requisite for its deposition, since it is never suspended in the air, like fog. Plants derive a great part of their nourishment from this source; and as each possesses a power of radiation peculiar to itself, they are capable of procuring a sufficient supply for their wants.

Rain is formed by the mixing of two masses of air of different temperatures; the colder part, by abstracting from the other the heat which holds it in solution, occasions the particles to approach each other and form drops of water, which, becoming too heavy to be sustained by the atmosphere, sink to the earth by gravitation in the form of rain. The contact of two strata of air of different temperatures, moving rapidly in opposite directions, occasions an abundant precipitation of rain.

An accumulation of caloric invariably produces light: with the exception of the gases, all bodies which can endure the requisite degree of heat without decomposition begin to emit light at the same temperature; but when the quantity of caloric is so great as to render the affinity of their component particles less than their affinity for the oxygen of the atmosphere, a chemical combination takes place with the oxygen, light and heat are evolved, and fire is produced. Combustion—so essential for our comfort, and even existence—takes place very easily from the small affinity between the component parts of atmospheric air, the oxygen being nearly in a free state; but as the cohesive force of the particles of different substances is very variable, different degrees of heat are requisite to produce their combustion. The tendency of heat to a state of equal diffusion or equilibrium, either by radiation or contact, makes it necessary that the chemical combination which occasions combustion should take place instantaneously; for if the heat were developed progressively, it would be dissipated by degrees, and would never accumulate sufficiently to produce a temperature high enough for the evolution of flame.

Though it is a general law that all bodies expand by heat and contract by cold, yet the absolute change depends upon the nature of the substance. Gases expand more than liquids, and liquids more than solids. The expansion of air is more than eight times that of water, and the increase in the bulk of water is at least forty-five times greater than that of iron. The expansion of solids and liquids increases uniformly with the temperature, between certain limits, this change of bulk, corresponding to the variation of heat, is one of the most important of its effects, since it furnishes the means of measuring relative temperature by the thermometer and pyrometer. The expansive force of caloric has a constant tendency to overcome the attraction of cohesion, and to separate the constituent particles of solids and fluids; by this separation the attraction of aggregation is more and more weakened, till at last it is entirely overcome, or even changed into repulsion. By the continual addition of caloric, solids may be made to pass into liquids, and from liquids to the aëriform state, the dilatation increasing with the temperature; but every substance expands according to a law of its own. Metals dilate uniformly from the freezing to the boiling points of the thermometer; the uniform expansion of the gases extends between still wider limits; but as liquidity is a state of transition from the solid to the aëriform condition, the equable dilatation of liquids has not so extensive a range. The rate of expansion of solids varies at their transition to liquidity, and that of liquids is no longer equable near their change to an aëriform state. There are exceptions, however, to the general laws of expansion; some liquids have a maximum density corresponding to a certain temperature, and dilate whether that temperature be increased or diminished. For example,—water expands whether it be heated above or cooled below 40°. The solidification of some liquids, and especially their crystallization, is always accompanied by an increase of bulk. Water dilates rapidly when converted into ice, and with a force sufficient to split the hardest substances. The formation of ice is therefore a powerful agent in the disintegration and decomposition of rocks, operating as one of the most efficient causes of local changes in the structure of the crust of the earth, of which we have experience in the tremendous *éboulemens* of mountains in Switzerland.

Heat is propagated with more or less rapidity through all bodies; air is the worst conductor, and consequently mitigates the severity of cold climates by preserving the heat imparted to the earth by the sun. On the contrary, dense bodies, especially metals, possess the power of conduction in the greatest degree, but the transmission requires time. If a bar of iron, twenty inches long, be heated at one extremity, the caloric takes four minutes in passing to the other. The particle of the metal that is first heated communicates its caloric to the second, and the second to the third; so that the temperature of the intermediate molecule at any instant is increased by the excess of the temperature of the first above its own, and diminished by the excess of its own temperature above that of the third. That, however, will not be the temperature indicated by the thermometer, because, as soon as the particle is more heated than the surrounding atmosphere, it will lose its caloric by radiation, in proportion to the excess of its actual temperature above that of the air.

The velocity of the discharge is directly proportional to the temperature, and inversely as the length of the bar. As there are perpetual variations in the temperature of all terrestrial substances, and of the atmosphere, from the rotation of the earth and its revolution round the sun, from combustion, friction, fermentation, electricity, and an infinity of other causes, the tendency to restore the equability of temperature by the transmission of caloric must maintain all the particles of matter in a state of perpetual oscillation, which will be more or less rapid according to the conducting powers of the substances. From the motion of the heavenly bodies about their axes, and also round the sun, exposing them to perpetual changes of temperature, it may be inferred that similar causes will produce like effects in them too. The revolutions of the double stars show that they are not at rest, and though we are totally ignorant of the changes that may be going on in the nebulæ and millions of other remote bodies, it is more than probable that they are not in absolute repose; so that, as far as our knowledge extends, motion seems to be a law of matter.

Heat applied to the surface of a fluid is propagated downwards very slowly, the warmer, and consequently lighter strata always remaining at the top. This is the reason why the water at the bottom of lakes fed from alpine chains is so cold; for the heat of the sun is transfused but a little way below the surface. When heat is applied below a liquid, the particles continually rise as they become specifically lighter, in consequence of the caloric, and diffuse it through the mass, their place being perpetually supplied by those that are more dense. The power of conducting heat varies materially in different liquids. Mercury conducts twice as fast as an equal bulk of water, which is the reason why it appears to be so cold. A hot body diffuses its caloric in the air by a double process. The air in contact with it, being heated, and becoming lighter, ascends and scatters its caloric, while at the same time another portion is discharged in straight lines by the radiating powers of the surface. Hence a substance cools more rapidly in air than in vacuo, because in the latter case the process is carried on by radiation alone. It is probable that the earth, having originally been of very high temperature, has become cooler by radiation only. The ethereal medium must be too rare to carry off much caloric.

Besides the degree of heat indicated by the thermometer, caloric pervades bodies in an imperceptible or latent state; and their capacity for heat is so various, that very different quantities of caloric are required to raise different substances to the same sensible temperature; it is therefore evident that much of the caloric is absorbed, or latent and insensible to the thermometer. The portion of caloric requisite to raise a body to a given temperature is its specific heat; but latent heat is that portion of caloric which is employed in changing the state of bodies from solid to liquid, and from liquid to vapour. When a solid is converted into a liquid, a greater quantity of caloric enters into it than can be detected by the thermometer; this accession of caloric does not make the body warmer, though it converts it into a liquid, and is the principal cause of its fluidity. Ice remains at the temperature of 32° of Fahrenheit till it has combined with or absorbed 140° of caloric, and then it melts, but without raising the temperature of the water above 32°; so that water is a compound of ice and caloric. On the contrary, when a liquid is converted into a solid, a quantity of caloric leaves it without any diminution of its temperature. Water at the temperature of 32° must part with 140° of caloric before it freezes. The slowness with which water freezes, or ice thaws, is a consequence of the time required to give out or absorb 140° of latent heat. A considerable degree of cold is often felt during a thaw, because the ice, in its transition from a solid to a liquid state, absorbs sensible heat from the atmosphere and other bodies, and, by rendering it latent, maintains them at the temperature of 32° while melting. According to the same principle, vapour is a combination of caloric with a liquid. About 1000° of latent heat exists in steam without raising its temperature: that is, boiling water, at the temperature of 212°, must absorb about 1000° of caloric before it becomes steam; and steam at 212° must part with the same quantity of latent caloric when condensed into water. The elasticity of steam may be increased to an enormous degree by increasing its temperature under pressure, yet its latent heat remains the same; however, it acquires an additional quantity, if allowed to expand; so that the latent heat of high-pressure steam issuing from a boiler is really two-fold—the latent heat of elastic fluidity and that of expansion. High-pressure steam expands the instant it comes into the air; the latent heat of expansion is increased at the expense of the latent heat of fluidity, in consequence of which, a portion of the steam is instantly condensed, and then the remaining portion, being mixed with air and particles of water, is so much reduced in temperature, that the hand may be plunged, without injury, into high-pressure steam, the instant it issues from the orifice of a boiler.

The latent heat of air, and of all elastic fluids, may be forced out by sudden compression, like squeezing water out of a sponge. The quantity of heat brought into action in this way is very well illustrated in the experiment of igniting a piece of tinder by the sudden compression of air by a piston thrust into a cylinder closed at one end: the developement of heat on a stupendous scale is exhibited in lightning, which is produced by the violent compression of the atmosphere during the passage of the electric fluid. Prodigious quantities of heat are constantly becoming latent, or are disengaged by the changes of condition to which substances are liable in passing from the solid to the liquid, and from the liquid to the gaseous form, or the contrary, occasioning endless vicissitudes of temperature over the globe.

The application of heat to the various branches of the mechanical and chemical arts has, within a few years, effected a greater change in the condition of man than had been accomplished in any equal period of his existence. Armed by the expansion and condensation of fluids with a power equal to that of the lightning itself, conquering time and space, he flies over plains, and travels on paths cut by human industry even through mountains, with a velocity and smoothness more like planetary than terrestrial motion; he crosses the deep in opposition to wind and tide; by releasing the strain on the cable, he rides at anchor fearless of the storm; he makes the elements of air and water the carriers of warmth, not only to banish winter from his home, but to adorn it even during the snow-storm with the blossoms of spring; and like a magician, he raises from the gloomy and deep abyss of the mine, the spirit of light to dispel the midnight darkness.

It has been observed that heat, like light and sound, probably consists in the undulations of an elastic medium. All the principal phenomena of heat may actually be illustrated by a comparison with those of sound. The excitation of heat and sound are not only similar, but often identical, as in friction and percussion; they are both communicated by contact and radiation; and Dr. Young observes, that the effect of radiant heat in raising the temperature of a body upon which it falls resembles the sympathetic agitation of a string, when the sound of another string, which is in unison with it, is transmitted to it through the air. Light, heat, sound, and the waves of fluids, are all subject to the same laws of reflection, and, indeed, their undulatory theories are perfectly similar. If, therefore, we may judge from analogy, the undulations of some of the heat-producing rays must be less frequent than those of the extreme red of the solar spectrum; but if the analogy were perfect, the interference of two hot rays ought to produce cold, since darkness results from the interference of two undulations of light, silence ensues from the interference of two undulations of sound; and still water, or no tide, is the consequence of the interference of two tides. The propagation of sound, however, requires a much denser medium than that either of light or heat, its intensity diminishes as the rarity of the air increases; so that at a very small height above the surface of the earth, the noise of the tempest ceases, and the thunder is heard no more in those boundless regions where the heavenly bodies accomplish their periods in eternal and sublime silence.

A consciousness of the fallacy of our judgment is one of the most important consequences of the study of nature. This study teaches us that no object is seen by us in its true place, owing to aberration; that the colours of substances are solely the effects of the action of matter upon light, and that light itself, as well as heat and sound, are not real beings, but mere modes of action communicated to our perceptions by the nerves. The human frame may therefore be regarded as an elastic system, the different parts of which are capable of receiving the tremors of elastic media, and of vibrating in unison with any number of superposed undulations, all of which have their perfect and independent effect. Here our knowledge ends; the mysterious influence of matter on mind will in all probability be for ever hid from man.

SECTION XXVI.

The sun and some of the planets appear to be surrounded with atmospheres of considerable density. According to the observations of Schröeter, the atmosphere of Ceres is more than 668 miles high, and that of Pallas has an elevation of 465 miles. It is remarkable that not a trace of atmosphere can be perceived in Vesta, and that Jupiter, Saturn, and Mars, have very little. The attraction of the earth has probably deprived the moon of hers, for the refractive power of the air at the surface of the earth is at least a thousand times as great as the refraction at the surface of the moon. The lunar atmosphere, therefore, must be of a greater degree of rarity than can be produced by our best air-pumps; consequently no terrestrial animal could exist in it.

What the body of the sun may be, it is impossible to conjecture; but he seems to be surrounded by a mottled ocean of flame, through which his dark nucleus appears like black spots, often of enormous size. These spots are almost always comprised within a zone of the sun's surface, whose breadth, measured on a solar meridian, does not extend beyond $30\frac{1}{2}°$ on each side of his equator, though they have been seen at the distance of $39\frac{1}{2}°$. From their extensive and rapid changes, there is every reason to suppose that the exterior and incandescent part of the sun is gaseous. The solar rays probably arising from chemical processes that continually take place at his surface are transmitted through space in all directions; but notwithstanding the sun's magnitude, and the inconceivable heat that must exist at his surface, as the intensity both of his light and heat diminishes as the square of the distance increases, his kindly influence can hardly be felt at the boundaries of our system. The power of the solar rays depends much upon the manner in which they fall, as we readily perceive from the different climates on our globe. In winter the earth is nearer the sun by about a thirtieth than in summer, but the rays strike the northern hemisphere more obliquely in winter than in the other half of the year. In Uranus the sun must be seen like a small but brilliant star, not above the hundred and fiftieth part so bright as he appears to us; but that is 2000 times brighter than our moon to us, so that he really is a sun to

Uranus, and probably imparts some degree of warmth. But if we consider that water would not remain fluid in any part of Mars, even at his equator, and that in the temperate zones of the same planet even alcohol and quicksilver would freeze, we may form some idea of the cold that must reign in Uranus, though it cannot exceed that of the surrounding space.

It is found by experience that heat is developed in opaque and translucent substances by their absorption of solar light, but that the sun's rays do not alter the temperature of perfectly transparent bodies through which they pass. As the temperature of the pellucid planetary space cannot be affected by the passage of the sun's light and heat, neither can it be raised by the heat radiated from the earth, consequently its temperature must be invariable. The atmosphere, on the contrary, gradually increasing in density towards the surface of the earth, becomes less pellucid, and therefore gradually increases in temperature both from the direct action of the sun, and from the radiation of the earth. Lambert had proved that the capacity of the atmosphere for heat varies according to the same law with its capacity for absorbing a ray of light passing through it from the zenith, whence M. Svanberg found that the temperature of space is 58° below the zero point of Fahrenheit's thermometer; and from other researches, founded upon the rate and quantity of atmospheric refraction, he obtained a result which only differs from the preceding by half a degree. M. Fourier has arrived at nearly the same conclusion, from the law of the radiation of the heat of the terrestrial spheroid, on the hypothesis of its having nearly attained its limit of temperature in cooling down from its supposed primitive state of fusion. The difference in the result of these three methods, totally independent of one another, only amounts to the fraction of a degree. Thus, as the temperature of space is uniform, it follows that no part of Uranus can experience more than 90° of cold, which only exceeds that which Sir Edward Parry suffered during one day at Melville Island, by 3°.

The climate of Venus more nearly resembles that of the earth, though, excepting perhaps at her poles, much too hot for animal and vegetable life as they exist here: but in Mercury, the mean heat, arising only from the intensity of the sun's rays, must be above that of boiling quicksilver, and water would boil even at his poles. Thus the planets, though kindred with the earth in motion and structure, are totally unfit for the habitation of such a being as man.

The direct light of the sun has been estimated

to be equal to that of 5563 wax candles of moderate size, supposed to be placed at the distance of one foot from the object: that of the moon is probably only equal to the light of one candle at the distance of twelve feet; consequently the light of the sun is more than three hundred thousand times greater than that of the moon; for which reason the light of the moon either imparts no heat, or it is too feeble to penetrate the glass of the thermometer, even when brought to a focus by a mirror. The intensity of the sun's light diminishes from the centre to the circumference of the solar disc; but in the moon the gradation is reversed.

Much has been done within a few years to ascertain the manner in which heat is distributed over the surface of our planet, and the variations of climate; which in a general view mean every change of the atmosphere, such as of temperature, humidity, variations of barometric pressure, purity of air, the serenity of the heavens, the effects of winds, and electric tension. Temperature depends upon the property which all bodies possess, more or less, of perpetually absorbing and emitting or radiating heat. When the interchange is equal, the temperature of a body remains the same; but when the radiation exceeds the absorption, it becomes colder, and *vice versâ*. But in order to determine the distribution of heat over the surface of the earth, it is necessary to find a standard by which the temperature in different latitudes may be compared. For that purpose it is requisite to ascertain by experiment the mean temperature of the day, of the month, and of the year, at as many places as possible throughout the earth. The annual average temperature may be found by adding the mean temperatures of all the months in the year, and dividing the sum by twelve. The average of ten or fifteen years will give it with tolerable accuracy; for although the temperature in any place may be subject to very great variations, yet it never deviates more than a few degrees from its mean state, which consequently offers good standard of comparison.

If climate depended solely upon the heat of the sun, all places having the same latitude would have the same mean annual temperature. The motion of the sun in the ecliptic, indeed, occasions perpetual variations in the length of the day, and in the direction of the rays with regard to the earth; yet, as the cause is periodic, the mean annual temperature from the sun's motion alone must be constant in each parallel of latitude. For it is evident that the accumulation of heat in the long days of summer, which is but little diminished by radiation during the short nights, is balanced by the small quantity of heat received during the short days in winter and its radiation in the long frosty and clear nights. In fact, if the globe were everywhere on a level with the surface of the sea, and also of the same substance, so as to absorb heat equally, and radiate the same, the mean heat of the sun would be regularly distributed over its surface in zones of equal annual temperature parallel to the equator, from which it would decrease to each pole as the square of the cosine of the latitude; and its quantity would only depend upon the altitudes of the sun, atmospheric currents, and the internal heat of the earth evinced by the vast number of volcanos and hot springs, in every region from the equator to the polar circles, which has probably been cooling down to its present state for thousands of ages. The distribution of heat, however, in the same parallel is very irregular in all latitudes, except between the tropics, where the isothermal lines, or the lines passing through places of equal mean annual temperature, are parallel to the equator. The causes of disturbance are very numerous; but such as have the greatest influence, according to Humboldt, to whom we are indebted for the greater part of what is known on the subject, are the elevation of the continents, the distribution of land and water over the surface of the globe, exposing different absorbing and radiating powers; the variations in the surface of the land, as forests, sandy deserts, verdant plains, rocks, &c., mountain-chains covered with masses of snow, which diminish the temperature; the reverberation of the sun's rays in the valleys, which increases it; and the interchange of currents, both of air and water, which mitigate the rigour of climates; the warm currents from the equator softening the severity of the polar frosts, and the cold currents from the poles tempering the intense heat of the equatorial regions. To these may be added cultivation, though its influence extends over but a small portion of the globe, only a fourth part of the land being inhabited.

Temperature does not vary so much with latitude as with the height above the level of the sea; and the decrease is more rapid in the higher strata of the atmosphere than in the lower, because they are farther removed from the radiation of the earth, and being highly rarefied, the heat is diffused through a larger space. A portion of air at the surface of the earth, whose temperature is 70° of Fahrenheit, if carried to the height of two miles and a half, will expand so much that its temperature will be reduced 50°; and in the ethereal regions the temperature is 90° below the point of congelation.

The height at which snow lies perpetually decreases from the equator to the poles, and is higher in summer than in winter; but it varies from many circumstances. Snow rarely falls when the cold is intense and the atmosphere dry. Extensive forests produce moisture by their evaporation, and high table-lands, on the contrary, dry and warm the air. In the Cordilleras of the Andes, plains of only twenty-five square leagues raise the temperature as much as three or four degrees above what is found at the same altitude on the rapid declivity of a mountain, consequently the line of perpetual snow varies according as one or other of these causes prevails. Aspect has also a great influence; the line of perpetual snow is much more elevated on the southern than on the northern side of the Himalaya mountains; but on the whole it appears that the mean height between the tropics at which the snow lies perpetually is about 15207 feet above the level of the sea; whereas snow does not cover the ground continually at the level of the sea till near the north pole. In the southern hemisphere, however, the cold is greater than in the northern. In Sandwich land, between the 54th and 58th degrees of latitude, perpetual snow and ice extend to the sea-beach; and in the island of St. George's, in the 53rd degree of south latitude, which corresponds with the latitude of the central counties of England, perpetual snow descends even to the level of the ocean. This preponderance of cold in the southern hemisphere cannot be altogether attributed to the winter being longer than ours by so small a quantity as 7¾ days, even allowing to that its due influence; but it is probably owing to the open sea round the south pole, which permits the icebergs to descend to a lower latitude by ten degrees than they do in the northern hemisphere, on account of the numerous obstructions opposed to them by the islands and continents about the north pole. Icebergs seldom float farther to the south than the Azores; whereas those that come from the south pole descend as far as the Cape of Good Hope, and occasion a continual absorption of heat in melting.

The influence of mountain-chains does not wholly depend upon the line of perpetual congelation; they attract and condense the vapours floating in the air, and send them down in torrents of rain; they radiate heat into the atmosphere at a lower elevation, and increase the temperature of the valleys by the reflection of the sun's rays, and by the shelter they afford against prevailing winds. But, on the contrary, one of the most general and powerful causes of cold arising from the vicinity of mountains is the freezing currents of wind which rush from their lofty peaks along the rapid declivities, chilling the surrounding valleys: such is the cutting north wind called the bise in Switzerland.

Next to elevation, the difference in the radiating and absorbing powers of the sea and land has the greatest influence in disturbing the regular distribution of heat. The extent of the dry land is not above the fourth part of that of the ocean, so that the general temperature of the atmosphere, regarded as the result of the partial temperatures of the whole surface of the globe, is most powerfully modified by the sea; besides, the ocean acts more uniformly on the atmosphere than the diversified surface of the solid mass does, both by the equality of its curvature and its homogeneity. In opaque substances the accumulation of heat is confined to the stratum nearest the surface; but the seas become less heated at their surface than the land, because the solar rays, before being extinguished, penetrate the transparent liquid to a greater depth, and in greater numbers than in the opaque masses. On the other hand, water has a considerable radiating power, which, together with evaporation, would reduce the surface of the ocean to a very low temperature, if the cold particles did not sink to the bottom, on account of their superior density. The seas preserve a considerable portion of the heat they receive in summer, and, from their saltness, do not freeze so soon as fresh water: so that, in consequence of all these circumstances, the ocean is not subject to such variations of heat as the land; and, by imparting its temperature to the winds, it diminishes the intensity of climate on the coasts and in the islands, which are never subject to such extremes of heat and cold as are experienced in the interior of continents, though they are liable to fogs and rain from the evaporation of the adjacent seas. On each side of the equator, to the 48th degree of latitude, the surface of the ocean is in general warmer than the air above it: the mean of the difference of temperature at noon and midnight is about 1°·37, the greatest deviation never exceeding from 0°·36 to 2°·16, which is much cooler than the air over the land.

On land the temperature depends upon the nature of the soil and its products, its habitual moisture or dryness. From the eastern extremity of the Sahara desert quite across Africa, the soil is almost entirely barren sand, and the Sahara desert itself, without including Dafour or Dongola, extends over an area of 194000 square leagues, equal to twice the area of the Mediterranean Sea, and raises the temperature of the air by radiation from 90° to 100°, which must have a most extensive influence. On the contrary, vegetation cools the air by evaporation and the apparent radiation of cold from the leaves of plants, because they absorb more caloric than they give out. The graminiferous plains of South America cover an extent ten times greater than France, occupying no less than about 50000 square leagues, which is more than the whole chain of the Andes, and all the scattered mountain-groups of Brazil: these, together with the plains of North America and the steppes of Europe and Asia, must have an extensive cooling effect on the atmosphere, if it be considered that, in calm and serene nights, they cause the thermometer to descend 12° or 14°, and that, in the meadows and heaths in England, the absorption of heat by the grass is sufficient to cause the temperature to sink to the point of congelation during the night for ten months in the year. Forests cool the air also by shading the ground from the rays of the sun, and by evaporation from the boughs. Hales found that the leaves of a single plant of helianthus, three feet high, exposed nearly forty feet of surface; and if it be considered that the woody regions of the river Amazons, and the higher part of the Oroonoko, occupy an area of 260000 square leagues, some idea may be formed of the torrents of vapour which arise from the leaves of the forests all over the globe. However, the frigorific effects of their evaporation are counteracted in some measure by the perfect calm which reigns in the tropical wildernesses. The innumerable rivers, lakes, pools, and marshes interspersed through the continents absorb caloric, and cool the air by evaporation; but on account of the chilled and dense particles sinking to the bottom, deep water diminishes the cold of winter, so long as ice is not formed.

In consequence of the difference in the radiating and absorbing powers of the sea and land, their configuration greatly modifies the distribution of heat over the surface of the globe. Under the equator only one-sixth part of the circumference is land; and the superficial extent of land in the northern and southern hemispheres is in the proportion of three to one: the effect of this unequal division is greater in the temperate, than in the torrid zones, for the area of land in the northern temperate zone is to that in the southern as thirteen to one, whereas the proportion of land between the equator and each tropic is as five to four; and it is a curious fact, noticed by Mr. Gardner, that only one twenty-seventh part of the land of the globe has land diametrically opposite to it. This disproportionate arrangement of the solid part of the globe has a powerful influence on the temperature of the southern hemisphere. But, besides these greater modifications, the peninsulas, promontories, and capes, running out into the ocean, together with bays and internal seas, all affect the temperature: to these may be added, the position of continental masses with regard to the cardinal points. All these diversities of land and water affect the temperature by the agency of the winds. On this account the temperature is lower on the eastern coasts both of the New and Old World, than on the western; for, considering Europe as an island, the general temperature is mild in proportion as the aspect is open to the western ocean, the superficial temperature of which, as far north as the 45° and 50° of latitude, does not fall below 48° or 51° of Fahrenheit, even in middle of winter. On the contrary, the cold of Russia arises from its exposure to the northern and eastern winds; but the European part of that empire has a less rigorous climate than the Asiatic, because the whole northern extremity of Europe is separated from the polar ice by a zone of open sea, whose winter temperature is much above that of a continental country under the same latitude.

The interposition of the atmosphere modifies all the effects of the sun's heat; but the earth communicates its temperature so slowly, that M. Arago has occasionally found as much as from 14° to 18° of difference between the heat of the soil and that of the air two or three inches above it.

The circumstances which have been enumerated,

and many more, concur in disturbing the regular distribution of heat over the globe, and occasion numberless local irregularities: nevertheless the mean annual temperature becomes gradually lower from the equator to the poles; but the diminution of mean heat is most rapid between the 40° and 45° of latitude both in Europe and America, which accords perfectly with theory, whence it appears that the variation in the square of the cosine of the latitude which expresses the law of the change of temperature, is a maximum towards the 45° of latitude. The mean annual temperature under the line in Asia and America is about 81½° of Fahrenheit; in Africa it is said to be nearly 83°. The difference probably arises from the winds of Siberia and Canada, whose chilly influence is sensibly felt in Asia and America, even within 18° of the equator.

The isothermal lines are parallel to the equator, till about the 22° of latitude on each side of it, where they begin to lose their parallelism, and continue to do so more and more as the latitude augments. With regard to the northern hemisphere, the isothermal line of 59° of Fahrenheit passes between Rome and Florence, in latitude 43°; and near Raleigh, in North Carolina, latitude 36°; that of 50° of equal annual temperature runs through the Netherlands, latitude 51°;

and near Boston, in the United States, latitude 42½°; that of 41° passes near Stockholm, latitude 59½°; and St. George's Bay, Newfoundland, latitude 48°; and lastly, the line of 32°, the freezing point of water, passes between Ulea, in Lapland, latitude 66°, and Table Bay, on the coast of Labradore, latitude 54°.

Thus it appears, that the isothermal lines which are parallel to the equator for nearly 22°, afterwards deviate more and more; and from the observations of Sir Charles Giesecke in Greenland, of Mr. Scoresby in the Arctic seas, and also from those of Sir Edward Parry and Sir John Franklin, it is found that the isothermal lines of Europe and America entirely separate in the high latitudes, and surround two poles of maximum cold, one in America and the other in the north of Asia, neither of which coincides with the pole of the earth's rotation. These poles are both situate in about the eightieth parallel of north latitude; the Transatlantic pole is in the 100° of west longitude, about 5° to the north of Sir Graham Moore's Bay, in the Polar Seas, and the Asiatic pole is in the 95° of east longitude, a little to the north of the Bay of Taimura, near the North-East Cape. According to the estimation of Sir David Brewster, from the observations of M. de Humboldt and Captains Parry and Scoresby, the mean annual temperature of the Asiatic pole is nearly 1° of Fahrenheit's thermometer, and that of the transatlantic pole about $3\frac{1}{2}$° below zero, whereas he supposes the mean annual temperature of the pole of rotation to be 4° or 5°. It is believed that two corresponding poles of maximum cold exist in the southern hemisphere, though observations are wanting to trace the course of the southern isothermal lines with the same accuracy as the northern.

The isothermal lines, or such as pass through places where the mean annual temperature of the air is the same, do not always coincide with the isogeothermal lines, which are those passing through places where the mean temperature of the ground is the same. The mean heat of the earth is determined from that of springs, and if the spring be on elevated ground, the temperature is reduced by computation to what it would be at the level of the sea, assuming that the heat of the soil varies according to the same law as the heat of the atmosphere, which is about a degree of Fahrenheit's thermometer for every 656 feet. From a comparison of the temperature of numerous springs with that of the air, Sir David Brewster concludes that there is a particular line passing nearly through Berlin, at which the temperature of springs and that of the atmosphere coincide; that in approaching the Arctic Circle the temperature of springs is always higher than that of the air, while proceeding towards the equator it is lower. He likewise found that the isogeothermal lines are always parallel to the isothermal lines, consequently the same general formulæ will serve to determine both, since the difference is a constant quantity, obtained by observation, and depending upon the distance of the place from the neutral isothermal line. These results are confirmed by the observations of M. Kupffer, of Kasan, during his excursions to the north, which show that the European and the American portions of the isogeothermal line of 32° Fahrenheit actually separate, and go round the two poles of maximum cold. This traveller remarked also, that the temperature both of the air and of the soil decreases most rapidly towards the 45° of latitude. The temperature of the ground at the equator is lower on the coasts and islands than in the interior of the continents; the warmest part is in the interior of Africa, but the temperature is obviously affected by the nature of the soil, especially if it be volcanic.

It is evident that places may have the same mean annual temperature, and yet differ materially in climate. In one the winters may be mild and the summers cool: whereas another may experience the extremes of heat and cold. Lines passing through places having the same mean summer or winter temperature, are neither parallel to the isothermal, the geothermal lines, nor to one another, and they differ still more from the parallels of latitude. In Europe, the latitude of two places which have the same annual heat never differs more than 8° or 9°; whereas the difference in the latitude of those having the same mean winter temperature is sometimes as much as 18° or 19°. At Kasan, in the interior of Russia, in latitude 55°·48, nearly the same with that of Edinburgh, the mean annual temperature is about 37°·6; at Edinburgh it is 47°·84. At Kasan, the mean summer temperature is 64°·84, and that of winter 2°·12, whereas at Edinburgh the mean summer temperature is 58°·28, and that of winter 38°·66. Whence it appears that the difference of winter temperature is much greater than that of the summer. At Quebec, the summers are as warm as those in Paris, and grapes sometimes ripen in the open air; whereas the winters are as severe as in Petersburg; the snow lies five feet deep for several months, wheel-carriages cannot be used, the ice is too hard for skating, travelling is performed in sledges, and frequently on the ice of the river St. Lawrence. The cold at Melville Island, on the 15th of January, 1820, according to Sir Edward Parry, was 55° below the zero of Fahrenheit's thermometer, only 3° above the temperature of the ethereal regions, yet the summer heat in these high latitudes is insupportable.

SECTION XXVII.

The gradual decrease of temperature in the air and in the earth, from the equator to the poles, is clearly indicated by its influence on vegetation. In the valleys of the torrid zone, where the mean annual temperature is very high, and where there is abundance of moisture, nature adorns the soil with all the luxuriance of perpetual summer. The palm, the bombax ceiba, and a variety of magnificent trees, tower to the height of a hundred and fifty or two hundred feet above the banana, the bamboo, the arborescent fern, and numberless other tropical productions, so interlaced by creeping and parasitical plants, as often to present an impenetrable barrier. But the richness of vegetation gradually diminishes with the temperature; the splendour of the tropical forest is succeeded by the regions of the olive and vine; these again yield to the verdant meadows of more temperate climes; then follow the birch and the pine, which probably owe their existence in very high latitudes more to the warmth of the soil than to that of the air; but even these enduring plants become dwarfish, stunted shrubs, till a verdant carpet of mosses and lichens, enamelled with flowers, exhibits the last signs of vegetable life during the short but fervent summers at the polar regions. Such is the effect of cold on the vegetable kingdom, that the numbers of species growing under the line and in the northern latitudes of 45° and 68°, are in the proportion of the numbers 12, 4, and 1. But notwithstanding the remarkable difference between a tropical and polar Flora, moisture seems to be almost the only requisite for vegetation, since neither heat, cold, nor even darkness destroy the fertility of nature; in salt plains and sandy deserts alone hopeless barrenness prevails. Plants grow on the borders of hot springs —they form the oases, wherever moisture exists, among the burning sands of Africa—they are found in caverns void of light, though generally blanched and feeble—the ocean teems with vegetation—the snow itself not only produces a red alga, discovered by Saussure in the frozen declivities of the Alps, found in abundance by the author crossing the Col de Bonhomme from Savoy to Piedmont, and by the polar navigators in the arctic regions, but it affords shelter to the productions of those inhospitable climes, against the piercing winds that sweep over fields of everlasting ice. Those interesting mariners narrate that, under this cold defence, plants spring up, dissolve the snow a few inches round, and that the part above, being again quickly frozen into a transparent sheet of ice, admits the sun's rays, which warm and cherish the plant in this natural hot-house, till the returning summer renders such protection unnecessary.

By far the greater part of the hundred and ten thousand known species of plants are indigenous in equinoctial America; Europe contains about half the number; Asia with its islands somewhat less than Europe; New Holland, with the islands in the Pacific, still less; and in Africa there are fewer vegetable productions than in any part of the globe, of equal extent. Very few social plants, such as grasses and heaths that cover large tracts of land, are to be found between the tropics, except on the sea coasts, and elevated plains. In the equatorial regions, where the heat is always great, the distribution of plants depends upon the mean annual temperature; whereas in temperate zones the distribution is regulated in some degree by the summer heat. Some plants require a gentle warmth of long continuance, others flourish most where the extremes of heat and cold are greater. The range of wheat is very great: it may be cultivated as far north as the 60° of latitude, but in the torrid zone, it will seldom form an ear below an elevation of 4500 feet above the level of the sea from the exuberance of vegetation; nor will it ripen above the height of 10800 feet, though much depends upon local circumstances. The best wines are produced between the 30° and 45° of north latitude. But with regard to the vegetable kingdom, elevation is equivalent to latitude, as far as temperature is concerned. In ascending the mountains of the torrid zone, the richness of the tropical vegetation diminishes with the height; a succession of plants similar, though not identical with those found in latitudes of corresponding mean temperature takes place; the lofty forests lose by degrees their splendour, stunted shrubs succeed, till at last the progress of the lichen is checked by eternal snow. On the volcano of Teneriffe, there are five successive zones, each producing a distinct race of plants. The first is the region of vines, the next that of laurels, these are followed by the districts of pines, of mountain broom, and of grass; the whole covering the declivity of the peak through an extent of 11200 feet of perpendicular height.

Near the equator the oak flourishes at the height of 9200 feet above the level of the sea, and on the lofty range of the Hymalaya the primula, the convallaria, and the veronica blossom, but not the primrose, the lily of the valley, or the veronica which adorn our meadows; for although the herbarium collected by Mr. Moorcroft on his route from Neetee to Daba and Garlope in Chinese Tartary, at elevations as high or even higher than Montblanc, abounds in Alpine and European genera, the species are universally different, with the single exception of the rhodiola rosea, which is identical with the species that blooms in Scotland. It is not in this instance alone that similarity of climate obtains without identity of productions; throughout the whole globe, a certain analogy both of structure and appearance is frequently discovered between plants under corresponding circumstances, which are yet specifically different. It is even said, that a distance of 25° of latitude occasions a total change not only of vegetable productions, but of organised beings. Certain it is, that each separate region both of land and water, from the frozen shores of the polar circles, to the burning regions of the torrid zone, possesses a flora of species peculiarly its own. The whole globe has been divided by botanical geographers into twenty-seven botanical districts, differing almost entirely in their specific vegetable productions; the limits of which are most decided when they are separated by a wide expanse of ocean, mountain chains, sandy deserts, salt plains, or internal seas. A considerable number of plants are common to the northern regions of Asia, Europe, and America, where these continents almost unite; but in approaching the south, the floras of these three great divisions of the globe differ more and more even in the same parallels of latitude, which shows that temperature alone is not the cause of the almost complete diversity of species that everywhere prevails. The floras of China, Siberia, Tartary, of the European district including central Europe and the coasts of the Mediterranean, and the oriental region, comprising the countries round the Black and Caspian Seas, all differ in specific character. Only twenty-four species were found by MM. Bonpland and Humboldt in equinoctial America, that are identical with those of the Old World; and Mr. Brown not only found that a peculiar vegetation exists in New Holland, between the thirty-third and thirty-fifth parallels of south latitude, but that, at the eastern and western extremities of these parallels, not one species is common to both, and that certain genera also are almost entirely confined to these spots. The number of species common to Australia and Europe are only 166 out of 4100, and probably some of these have been conveyed thither by the colonists. This proportion exceeds what is observed in southern Africa, and from what has been already stated, the proportion of European species in equinoctial America is still less.

Islands partake of the vegetation of the nearest

continents, but when very remote from land their floras are altogether peculiar. The Aleutian islands, extending between Asia and America, partake of the vegetation of the northern parts of both these continents, and may have served as a channel of communication. In Madeira and Teneriffe, the plants of Portugal, Spain, the Azores, and of the north coast of Africa are found, and the Canaries contain a great number of plants belonging to the African coast. But each of these islands possesses a flora that exists nowhere else, and St. Helena, standing alone in the midst of the Atlantic ocean, out of sixty-one indigenous species, produces only two or three recognised as belonging to any other part of the world.

It appears from the investigations of Humboldt that between the tropics the monocotyledonous plants, such as grasses and palms, which have only one seed-lobe, are to the dicotyledonous tribe, which have two seed-lobes, like most of the European species, in the proportion of one to four; in the temperate zones they are as one to six; and in the arctic regions, where mosses and lichens, which form the lowest order of the vegetable creation, abound, the proportion is as one to two. The annual monocotyledonous and dicotyledonous plants in the temperate zones amount to one-sixth of the whole, omitting the cryptogamia; in the torrid zone they scarcely form one-twentieth, and in Lapland one-thirtieth part. In approaching the equator, the ligneous exceed the number of herbaceous plants; in America, there are a hundred and twenty different species of forest-trees, whereas in the same latitude in Europe only thirty-four are to be found.

Similar laws appear to regulate the distribution of marine plants. M. Lamouroux has discovered that the groups of algæ affect particular temperatures or zones of latitude, though some few genera prevail throughout the ocean. The polar Atlantic basin, to the 40° of north latitude, presents a well-defined vegetation. The West Indian seas, including the gulf of Mexico, the eastern coast of South America, the Indian ocean and its gulfs, the shores of New Holland, and the neighbouring islands, have each their assemblage of distinct species. The Mediterranean possesses a vegetation peculiar to itself, extending to the Black Sea; and the species of marine plants on the coasts of Syria and in the port of Alexandria differ almost entirely from those of Suez and the Red Sea, notwithstanding the proximity of their geographical situation. It is observed that shallow seas have a different set of plants from such as are deeper and colder; and, like terrestrial vegetation, the algæ are most numerous towards the equator, where the quantity must be prodigious, if we may judge from the gulf-weed, which certainly has its origin in the tropical seas, and is drifted, though not by the gulf-stream, to higher latitudes, where it accumulates in such quantities, that the early Portuguese navigators, Columbus and Lerius, compared the sea to extensively inundated meadows, in which it actually impeded their ships and alarmed their sailors. Humboldt, in his Personal Narrative, mentions, that the most extensive bank of sea-weed is in the northern Atlantic, a little west of the meridian of Fayal, one of the Azores, between the 25° and 36° of latitude. Vessels returning to Europe from Monte Video, or from the Cape of Good Hope, cross this bank nearly at an equal distance from the Antilles and Canary islands. The other occupies a smaller space, between the 22° and 26° of north latitude, about eighty leagues west of the meridian of the Bahama islands, and is generally traversed by vessels on their passage from the Caicos to the Bermuda islands. These masses consist chiefly of one or two species of Sargassum, the most extensive genus of the order Fucoideæ.

Some of the sea-weeds grow to the enormous length of several hundred feet, and all are highly coloured, though many of them must grow in the deep caverns of the ocean in total, or almost total darkness; light, however, may not be the only principle on which the colour of vegetables depends, since Humboldt met with green plants growing in complete darkness at the bottom of one of the mines at Freuberg.

It appears that in the dark and tranquil caves of the ocean, on the shores alternately covered and deserted by the restless waves, on the lofty mountain and extended plain, in the chilly regions of the north, and in the genial warmth of the south, specific diversity is a general law of the vegetable kingdom, which cannot be accounted for by diversity of climate; and yet the similarity though not identity of species is such, under the same isothermal lines, that if the number of species belonging to one of the great families of plants be known in any part of the globe, the whole number of the phanerogamous or more perfect plants, and also the number of species composing the other vegetable families, may be estimated with considerable accuracy.

Various opinions have been formed on the original or primitive distribution of plants over the surface of the globe, but since botanical geography became a regular science, the phenomena observed have led to the conclusion that vegetable creation must have taken place in a number of distinctly different centres, each of which was the original seat of a certain number of peculiar species, which at first grew there and no where else. Heaths are exclusively confined to the old world, and no indigenous rose-tree has ever been discovered in the new; the whole southern hemisphere being destitute of that beautiful and fragrant plant. But this is still more confirmed by multitudes of particular plants having an entirely local and insulated existence, growing spontaneously in some particular spot and in no other place; as, for example, the cedar of Lebanon, which grows indigenously on that mountain and in no other part of the world.

The same laws obtain in the distribution of the animal creation. The zoophite, occupying the lowest place in animated nature, is widely scattered through the seas of the torrid zone, each species being confined to the district best fitted to its existence. Shell-fish decrease in size and beauty with their distance from the equator; and as far as is known, each sea has its own kind, and every basin of the ocean is inhabited by its peculiar tribe of fish. Indeed, MM. Peron and Le Sueur assert, that among the many thousands of marine animals which they had examined, there is not a single animal of the southern regions which is not distinguishable by essential characters from the analogous species in the northern seas. Reptiles are not exempt from the general law. The Saurian tribes of the four quarters of the globe differ in species, and although warm countries abound in venomous snakes, they are specifically different, and decrease both in the numbers and in the virulence of their poison with decrease of temperature. The dispersion of insects necessarily follows that of the vegetables which supply them with food, and in general it is observed, that each kind of plant is peopled by its peculiar inhabitants. Each species of bird has its particular haunt, notwithstanding the locomotive powers of the winged tribes. The emu is confined to Australia, the condor never leaves the Andes, nor the great eagle the Alps; and although some birds are common to every country, they are few in number. Quadrupeds are distributed in the same manner wherever man has not interfered. Such as are indigenous in one continent are not the same with their congeners in another; and with the exception of some kinds of bats, no warm-blooded animal is indigenous in the Polynesian Archipelago, nor in any of the islands on the borders of the central part of the Pacific.

In reviewing the infinite variety of organised beings that people the surface of the globe, nothing is more remarkable than the distinctions which characterise the different tribes of mankind, from the ebony skin of the torrid zone to the fair and ruddy complexion of Scandinavia, a difference which existed in the earliest recorded times, since the African is represented in the sacred writings to have been as black in the first ages of mankind as he is at the present day, and the most ancient Egyptian paintings confirm that truth; yet it appears from a comparison of the principal circumstances relating to the animal economy or physical character of the various tribes of mankind, that the different races are identical in species. Many attempts have been made to trace the various tribes back to a common origin, by collating the numerous languages which are, or have been, spoken. Some classes of these have few or no words in common, yet exhibit a remarkable analogy in the laws of their grammatical construction. The languages spoken by the native American nations afford examples of these; indeed the refinement in the grammatical construction of the tongues of the American savages leads to the belief that they must originally have been spoken by a much more civilized class of mankind. Some tongues have little or no resemblance in structure, though they correspond extensively in their vocabularies, as in the Syrian dialects. In all of these cases it may be inferred, that the nations speaking the languages in question are descended from the same stock; but the probability of a common origin is much greater in the Indo-European nations, whose languages, such as the Sanscrit, Greek, Latin, German, &c. have an affinity both in structure and correspondence of vocables. In many tongues not the smallest resemblance can be traced; length of time, however, may have obliterated the original identity. The conclusion drawn from the whole investigation is, that although the distribution of organized beings does not follow the direction of the isothermal lines, temperature has a very great influence on their physical development. Possibly, too, the nature of animated and inanimated creatures may be powerfully modified by the invisible agencies of electricity and magnetism, which probably pervade all the particles of matter; indeed the temperature of the air seems to be intimately connected with its electrical condition.

SECTION XXVIII.

Electricity is one of those imponderable agents pervading the earth and all substances, without affecting their volume or temperature, or even giving any visible sign of its existence when in a latent state, but when elicited, developing forces capable of producing the most sudden, violent, and destructive effects in some cases, while in others their action, though less energetic, is of indefinite and uninterrupted continuance. These modifications of the electric force, incidentally depending upon the manner in which it is excited, present phenomena of great diversity, but yet so connected as to justify the conclusion that they originate in a common principle.

Electricity may be called into activity by mechanical power, by chemical action, by heat, and by magnetic influence; but we are totally ignorant why it is roused from its neutral state by such means, or of the manner of its existence in bodies; whether it be a material agent, or merely a property of matter. However, as some hypothesis is necessary for explaining the phenomena observed, it is assumed to be a highly-elastic fluid, capable of moving with various degrees of facility through the pores or even the substance of matter; and as experience shows that bodies in one electric state attract, and in another repel each other, the hypothesis of two kinds, called positive and negative electricity, is adopted, but whether there really be two different fluids, or that the mutual attraction and repulsion of bodies arises from the redundancy and defect of their electricities, is of no consequence, since all the phenomena can be explained on either hypothesis. As each electricity has its peculiar properties, the science may be divided into branches, of which the following notice is intended to convey some idea.

Substances in which the positive and negative electricities are combined, being in a neutral state, neither attract nor repel; but there is a numerous class called electrics, in which the electric equilibrium is destroyed by friction: then the positive and negative electricities are called into action or separated; the positive is impelled in one direction, and the negative in another; those of the same kind repel, whereas those of different kinds attract each other. The attractive power is exactly equal to the repulsive force at equal distances, and when not opposed, they coalesce with great rapidity and violence, producing the electric flash, explosion, and shock; then equilibrium is restored, and the electricity remains latent till again called forth by a new exciting cause. One kind of electricity cannot be evolved without the evolution of an equal quantity of the opposite kind: thus, when a glass rod is rubbed with a piece of silk, as much positive electricity is elicited in the glass as there is negative in the silk. The kind of electricity depends more upon the mechanical condition than on the nature of the surface, for when two plates of glass, one polished and the other rough, are rubbed against each other, the polished surface acquires positive, and the rough negative electricity. The manner in which the friction is performed also alters the kind of electricity. Equal lengths of black and white ribbon, applied longitudinally to one another, and drawn between the finger and thumb, so as to rub their surfaces together, become electric; when separated, the black ribbon is found to have acquired negative electricity, and the white positive: but if the whole length of the black ribbon be drawn across the breadth of the white, the black will be positively, and the white negatively electric when separate. Electricity may be transferred from one body to another in the same manner as heat is communicated, and, like it too, the body loses by the transmission. Although no substance is altogether impervious to the electric fluid, nor is there any that does not oppose some resistance to its passage, yet it moves with much more facility through a certain class of substances called conductors, such as metals, water, the human body, &c., than through atmospheric air, glass, silk, &c., which are therefore called non-conductors; but the conducting power is affected both by temperature and moisture.

Bodies surrounded with non-conductors are said to be insulated, because, when charged, the electricity cannot escape; but when that is not the case, the electricity is conveyed to the earth, which is formed of conducting matter; consequently it is impossible to accumulate electricity in a conducting substance that is not insulated. There are a great many substances called non-electrics, in which electricity is not sensibly developed by friction, unless they be insulated, probably because it is carried off by their conducting power as soon as elicited. Metals, for example, which are said to be non-electrics, can be excited, but, being conductors, they cannot retain this state if in communication with the earth. It is probable that no bodies exist which are either perfect non-electrics or perfect non-conductors; but it is evident that electrics must be non-conductors to a certain degree, otherwise they could not retain their electric state.

It has been supposed that an insulated body remains at rest, because the tension of the electricity, or its pressure on the air which restrains it, is equal on all sides; but when a body in a similar state, and charged with the same kind of electricity, approaches it, that the mutual repulsion of the particles of the electric fluid diminishes the pressure of the fluid on the air on the adjacent sides of the two bodies, and increases it on their remote ends; consequently that equilibrium will be destroyed, and the bodies, yielding to the action of the preponderating force, will recede from or repel each other. When, on the contrary, they are charged with opposite electricities, it is alleged that the pressure upon the air on the adjacent sides will be increased by the mutual attraction of the particles of the electric fluid, and that on the further sides diminished; consequently that the force will urge the bodies towards one another, the motion in both cases corresponding to the forces producing it. An attempt has thus been made to attribute electrical attractions and repulsions to the mechanical pressure of the atmosphere; it is, however, more than doubtful whether these phenomena can be referred to that cause, but certain it is that, whatever the nature of these forces may be, they are not impeded in their action by the intervention of any substance whatever, provided it be not itself in an electric state.

A body charged with electricity, although perfectly insulated, so that all escape of electricity is precluded, tends to produce an electric state of the opposite kind in all bodies in its vicinity; positive electricity tends to produce negative electricity in a body near it, and *vice versâ*, the effect being greater as the distance diminishes. This power which electricity possesses of causing an opposite electrical state in its vicinity is called induction. When a body charged with either species of electricity is presented to a neutral one, its tendency, in consequence of the law of induction, is to disturb the electrical condition of the neutral body. The electrified body induces electricity contrary to its own in the adjacent part of the neutral one, and therefore an electrical state similar to its own in the remote part; hence the neutrality of the second body is destroyed by the action of the first, and the adjacent parts of the two, having now opposite electricities, will attract each other. The attraction between electrified and unelectrified substances is therefore merely a consequence of their altered state, resulting directly from the law of induction, and not an original law. The effects of induction depend upon the facility with which the equilibrium of the neutral state of a body can be overcome, a facility which is proportional to the conducting power of the body; consequently, the attraction exerted by an electrified substance upon another substance previously neutral will be much more energetic if the latter be a conductor than if it be a non-conductor.

The law of electrical attraction and repulsion has been determined by suspending a needle of gum lac horizontally by a silk fibre, the needle carrying at one end a piece of electrified gold-leaf. A globe charged with the same, or with the opposite kind of electricity, when presented to the gold-leaf, will repel or attract it, and will therefore cause the needle to vibrate more or less rapidly according to the distance of the globe. A comparison of the number of oscillations performed in a given time, at different distances, will determine the law of the variation of the electrical intensity, in the same manner that the force of gravitation is measured by the oscillations of the pendulum. Coulomb invented an instrument which balances the forces in question by the force of the torsion of a thread, which consequently measures their intensity. By this method he found that the intensity of the electrical attraction and repulsion varies inversely as the square of the distance. Since electricity can only be in equilibrio from the mutual repulsion of its particles,—which, according to these experiments, varies inversely as the square of the distance,—its distribution in different bodies depends upon the laws of mechanics, and therefore becomes a subject of analysis and calculation. The distribution of electricity has been so successfully determined by the analytical investigations of M. Poisson and Mr. Ivory, that all the computed phenomena have been confirmed by observation.

It is found by direct experiment that a metallic globe or cylinder contains the same quantity of electricity when hollow that it does when solid; therefore electricity is entirely confined to the surfaces of bodies, or, if it does penetrate their substance, the depth is inappreciable: consequently the quantity bodies are capable of receiving does not follow the proportion of their bulk, but depends principally upon the extent of surface over which it is spread; so that the exterior may be positively or negatively electric, while the interior is in a state of perfect neutrality.

Electricity of either kind may be accumulated to a great extent in insulated bodies, and as long as it is quiescent it occasions no sensible change in their properties, though it is spread over their surfaces in indefinitely thin layers. When restrained by the non-conducting power of the atmosphere, the tension or pressure exerted by the electric fluid against the air which opposes its escape is in the ratio compounded of the repulsive force of its own particles at the surface of the stratum of the fluid and of the thickness of that stratum; but as one of these elements is always proportional to the other, the total pressure on every point must be proportional to the square of the thickness. If this pressure be less than the coercive force of the air, the electricity is retained; but the instant it exceeds that force in any one point the electricity escapes, which it will do when the air is attenuated, or becomes saturated with moisture.

The power of retaining electricity depends also upon the shape of the body. It is most easily retained by a sphere, next to that by a spheroid, but it readily escapes from a point; and, on the contrary, a pointed object receives it with most facility. It appears from analysis that electricity, when in equilibrio, spreads itself in a thin stratum over the surface of a sphere, in consequence of the repulsion of its particles, which force is directed from the centre to the surface. In an oblong spheroid the intensity or thickness of the stratum of electricity at the extremities of the two axes is exactly in the proportion of the axes themselves; hence, when the ellipsoid is much elongated, the electricity becomes very feeble at the equator and powerful at the poles. A still greater difference in the intensities takes place in bodies of a cylindrical or prismatic form, and the more so in proportion as their length exceeds their breadth; therefore the electrical intensity is very powerful at a point, where nearly the whole electricity in the body will be concentrated.

A perfect conductor is not mechanically affected by the passage of electricity, if it be of sufficient size to carry off the whole; but it is shivered to pieces in an instant, if it be too small to carry off the charge: this also happens to a bad conductor. In that case the physical change is generally a separation of the particles, though it may occasionally be attributed to chemical action, or expansion from the heat evolved during the passage of the fluid; but all these effects are in proportion to the obstacles opposed to the freedom of its course. The heat produced by the electric shock is intense, fusing metals, and even volatilizing substances, though it is only accompanied by light when the fluid is obstructed in its passage. Electrical light is perfectly similar to solar light in its composition; it seems to arise from the condensation of the air, during the rapid motion of the electricity, and varies both in intensity and colour with the density of the atmosphere. Electricity is occasionally produced by pressure and fracture; several crystalline substances also become electric when heated, especially tourmaline, one end of which acquires positive, and the other negative electricity, while the intermediate part is neutral; but when broken through the middle each fragment is found to possess positive electricity at one end, and negative at the other, like the entire crystal. Electricity is evolved by bodies passing from a liquid to a solid state, also by the production and condensation of vapour, which is consequently a great source of atmospheric electricity.

The atmosphere, when clear, is almost always positively electric; its electricity is stronger in winter than in summer, during the day than in the night. The intensity increases for two or three hours from the time of sunrise, then decreases towards the middle of the day, and again augments as the sun declines, till about the time of sunset, after which it diminishes, and continues feeble during the night. Atmospheric electricity arises from an evolution of the electric fluid during the evaporation that is so abundant at the surface of the earth; and clouds probably owe their existence, or at least their form, to it, for they consist of hollow vesicles of vapour coated with electricity; as the electricity is either entirely positive or negative, the vesicles repel each other, which prevents them from uniting and falling down in rain. The friction of the surfaces of two strata of air moving in different directions, probably developes electricity; and if the strata be of different temperatures, a portion of the vapour they always contain will be deposited; the electricity evolved will be taken up by the vapour, and will cause it to assume the vesicular state constituting a cloud. A vast deal of electricity may be accumulated in this manner, which may either be positive or negative, and should two clouds charged with opposite kinds approach within a certain distance, the thickness of the coating of electricity will increase on the two sides of the clouds that are nearest to one another; and when the accumulation becomes so great as to overcome the coercive pressure of the atmosphere, a discharge takes place, which occasions a flash of lightning. The actual quantity of electricity in any one part of a cloud is extremely small; the intensity of the flash arises from the very great extent of surface occupied by the electricity, so that the clouds may be compared to enormous Leyden jars thinly coated with the electric fluid, which only acquires its intensity by its instantaneous condensation.

An interchange frequently takes place between the clouds and the earth, but on account of the extreme rapidity of lightning it is difficult to ascertain whether it goes from the clouds to the earth, or shoots upwards from the earth to the clouds, though there can be no doubt that it does both. M. Halvig measured the velocity of lightning by means of the camera lucida, and estimates that it is probably eight or ten miles in a second, or about forty times greater than that of sound; and M. Gay-Lussac has ascertained that a flash of lightning sometimes darts more than three miles at once in a straight line.

A person may be killed by lightning, although the explosion takes place at the distance of twenty miles, by what is called the back stroke. Suppose that the two extremities of a cloud highly charged with electricity hang down towards the earth, they will repel the electricity from the earth's surface, if it be of the same kind with their own, and will attract the other kind; and if a discharge should suddenly take place at one end of the cloud, the equilibrium will instantly be restored by a flash at that point of the earth which is under the other.

The pure air, at all times negatively electric, becomes intensely so on the approach of rain, snow, wind, hail, or sleet, but it afterwards varies on opposite sides, and the transitions are very rapid on the approach of a thunder-storm. An insulated conductor then gives out such quantities of sparks that it is dangerous to approach it, as was fatally experienced by Professor Richman, at Petersburg, who was struck dead by a globe of fire from the extremity of a conductor, while making experiments on atmospheric electricity. There is no instance on record of an electric cloud being dispelled by a conducting rod silently withdrawing the electric fluid; yet it may mitigate the stroke, or render it harmless if it should come. Sir John Leslie observes, that the efficacy of conductors depends upon the rapidity with which they transmit the electric energy; and as copper is found to transmit the fluid twenty times faster than iron, and as iron conducts it 400000000 times more rapidly than water, which conveys it several thousand times faster than dry stone, copper conductors afford the best protection, especially if they expose a broad surface, since the electric fluid is conveyed chiefly along the exterior of bodies. The object of a conductor being to carry off the electricity in case of a stroke, and not to invite an enemy, it ought to project very little, if at all, above the building.

The aurora borealis is decidedly an electrical phenomenon, which takes place in the highest regions of the atmosphere, since it is visible at the same time from places very far distant from each other. It is somehow connected with the magnetic poles of the earth, but it has never been seen so far north as the pole of the earth's rotation, nor does it extend to low latitudes. It generally appears in the form of a luminous arch, stretching more or less from east to west, but never from north to south; across the arch the coruscations are rapid, vivid, and of various colours. A similar phenomenon occurs in the high latitudes of the southern hemisphere. Mr. Faraday conjectures that the electric equilibrium of the earth is restored by means of the aurora conveying the electricity from the poles to the equator.

SECTION XXIX.

Galvanism is a peculiar kind of electricity, elicited by the force of chemical action, instead of friction. It is connected with one of the most brilliant periods of British science, from the splendid discoveries to which it led Sir Humphry Davy; but it has acquired additional interest since it has been proved, by the reciprocal action of galvanic and magnetic currents, that magnetism has no existence as a distinct or separate principle, but is only an effect of electricity: therefore, galvanism, as immediately connected with the theory of the earth and planets, forms a part of the physical account of their nature.

The disturbance of electric equilibrium, and a development of electricity, invariably accompanies the chemical action of a fluid on metallic substances, and is most plentiful when that action occasions oxidation. Metals vary in the quantity of electricity afforded by their combination with oxygen; but the greatest abundance is developed by the oxidation of zinc by weak sulphuric acid; and in conformity with the law, that one kind of electricity cannot be evolved without an equal quantity of the other being brought into activity, it is found that the acid is positively, and the zinc negatively electric. It has not yet been ascertained why equilibrium is not restored by the contact of these two substances, which are both conductors, and in opposite electrical states; however, the electrical and chemical changes are so connected, that unless the equilibrium be restored, the action of the acid will go on languidly, or stop as soon as a certain quantity of electricity is accumulated in the acid. The equilibrium, however, will be restored, and the action of the acid will be continuous, if a plate of copper be placed in contact with the zinc, both being partly immersed in the fluid; for the copper, not being acted upon by the acid, will serve as a conductor to convey the positive electricity from the acid to the zinc, and will at every instant restore the equilibrium, and then the oxidation of the zinc will go on rapidly. Thus three substances are concerned in forming a galvanic circuit, but it is indispensable that one of them be a fluid. The electricity so obtained will be very feeble, but it may be augmented by increasing the number of plates. In the common galvanic battery, the electricity which the fluid has acquired from the first plate of zinc exposed to its action, is taken up by the copper plate belonging to the second pair, and transferred to the second zinc plate with which it is connected. This second plate of zinc having thus acquired a larger portion of electricity than its natural share, communicates a larger quantity of electricity to the fluid in the second cell. This increased quantity is again transferred to the next pair of plates; and thus every succeeding alternation is productive of a further increase in the quantity of the electricity developed. This action, however, would stop unless a vent were given to the accumulated electricity, by establishing a communication between the positive and negative poles of the battery, by means of wires attached to the extreme plate at each end. When the wires are brought into contact, the galvanic circuit is completed, the electricities meet and neutralize each other, producing the shock and other electrical phenomena, and then the electric current continues to flow uninterruptedly in the circuit, as long as the chemical action lasts. The stream of positive electricity flows from the zinc to the copper, but as the battery ends in a zinc plate which communicates with the wire, the zinc end becomes the positive, and the copper the negative poles of a compound battery, which is exactly the reverse of what obtains in a single circuit.

Galvanic or voltaic, like common electricity, may either be considered to consist of two fluids passing in opposite directions through the circuit, the positive stream coming from the zinc, and the negative from the copper end of the battery; or, if the hypothesis of one fluid be adopted, the zinc end of the battery may be supposed to have an excess of electricity, and the copper end a deficiency.

Voltaic electricity is distinguished by two marked characters. Its intensity increases with the number of plates—its quantity with the extent of their surfaces. The most intense concentration of force is displayed by a numerous series of large plates, light and heat are copiously evolved, and chemical decomposition is accomplished with extraordinary energy; whereas, the electricity from one pair of plates is so feeble, whatever their size may be, that it gives no sign either of attraction or repulsion; and, even with a battery consisting of a very great number of plates, it is difficult to render the mutual attraction of its two wires sensible, though of opposite electricities.

The action of voltaic electricity differs materially from that of the ordinary kind. When a quantity of common electricity is accumulated, the restoration of equilibrium is attended by an instantaneous violent explosion, accompanied by the development of light, heat, and sound. The concentrated power of the fluid forces its way through every obstacle, disrupting and destroying the cohesion of the particles of the bodies through which it passes, and occasionally increasing its destructive effects by the conversion of fluids into steam from the intensity of the momentary heat, as when trees are torn to pieces by a stroke of lightning: even the vivid light which marks the path of the electric fluid is probably owing to the sudden compression of the air and other particles of matter during the rapidity of its passage; but the instant equilibrium is restored by this energetic action, the whole is at an end. On the contrary, when an accumulation takes place in a voltaic battery, equilibrium is restored the moment the circuit is completed; but so far is the electric stream from being exhausted, that it continues to flow silently and invisibly in an uninterrupted current supplied by a perpetual reproduction; and although its action on bodies is neither so sudden nor so intense as that of common electricity, yet it acquires such power from constant accumulation and continued action, that it ultimately surpasses the energy of the other. The two kinds of electricity differ in no circumstance more than in the development of heat. Instead of a momentary evolution, which seems to arise from a forcible compression of the particles of matter during the passage of the common electric fluid, the circulation of the voltaic electricity is accompanied by a continued development of heat, lasting as long as the circuit is complete, without producing either light or sound; and this appears to be its immediate direct effect, independent of mechanical action. Its intensity is greater than that of any heat that can be obtained by artificial means, so that it fuses substances which resist the action of the most powerful furnaces. The temperature of every part of a galvanic battery itself is raised during its activity.

When the battery is powerful, the luminous effects of galvanism are very brilliant; but considerable intensity is requisite to enable the electricity to force its way through the air on bringing the wires together from the opposite poles. Its transit is accompanied by light, and in consequence of the continuous supply of the fluid, sparks occur every time the contact of the wires is either broken or renewed. The most splendid artificial light known is produced by fixing pencils of charcoal at the extremities of the wires, and bringing them into contact. This light is the more remarkable as it appears to be independent of combustion, since the charcoal suffers no change, and likewise because it is equally vivid in such gases as do not contain oxygen. Though nearly as bright as solar light, it differs from it in possessing some of those rays of which the sunbeams are deficient, according to the experiments of M. Fraunhofer. Voltaic electricity is a powerful agent in chemical analysis; numerous instances might be given, but the decomposition of water is perhaps the most simple and elegant. Suppose a glass tube filled with very pure water, and corked at both ends: if one of the wires of an active galvanic battery be made to pass through one cork, and the other through the other cork, into the water, so that the extremities of the two wires shall be opposite and about a quarter of an inch asunder, chemical action will immediately take place, and gas will continue to rise from the extremities of both wires till the water has vanished. If an electric spark be then sent through the tube, the water will reappear. By arranging the experiment so as to have the gas given out by each wire separately, it is found that water consists of two parts of hydrogen and one of oxygen. The positive wire of the battery has a stronger affinity for oxygen than oxygen has for hydrogen; it consequently combines with the oxygen of the water, and sets the hydrogen free; but as the negative wire has a stronger affinity for hydrogen than hydrogen has for oxygen, it combines with the hydrogen of the water, and sets the oxygen free. If, therefore, an electric spark be sent through a mixture consisting of two parts of hydrogen and one of oxygen, the gases will combine and form water. The decomposition of the alkalies and earths by Sir Humphry Davy, and all chemical changes produced by the electric fluid, are accomplished on the same principle, and it appears that, in general, combustible substances go to the negative wire, while oxygen is evolved at the positive. The powerful efficacy of voltaic electricity in chemical decomposition arises from the continuance of its action, and its agency appears to be most exerted on fluids and substances which, by conveying the electricity partially and imperfectly, impede its progress. But it is now proved to be as efficacious in the composition as in the decomposition or analysis of bodies.

It had been observed that, when metallic solutions are subjected to galvanic action, a deposition of metal, generally in the form of minute crystals, takes place on the negative wire: by extending this principle, and employing a very feeble voltaic action, M. Becquerel has succeeded in forming crystals of a great proportion of the mineral substances precisely similar to those produced by nature. The electric state of metallic veins makes it possible that many natural crystals may have taken their form from the action of electricity bringing their ultimate particles, when in solution, within the narrow sphere of molecular attraction already mentioned as the great agent in the formation of solids. Both light and motion favour crystallization. Crystals which form in different liquids are generally more abundant on the side of the jar exposed to the light; and it is a well-known fact that still water, cooled below 32°, starts into crystals of ice the instant it is agitated. Light and motion are intimately connected with electricity, which may therefore have some influence on the laws of aggregation; this is the more likely, as a feeble action is alone necessary, provided it be continued for a sufficient time. Crystals formed rapidly are generally imperfect and soft, and M. Becquerel found that even years of constant voltaic action were necessary for the crystallization of some of the hard substances. If this law be general, how many ages may be required for the formation of a diamond!

Several fish possess the faculty of producing electrical effects. The most remarkable are the gymnotus electricus, found in South America, and the torpedo, a species of ray, frequent in the Mediterranean. The absolute quantity of electricity brought into circulation by the torpedo is so great that it effects the decomposition of water, has power sufficient to make magnets, and gives very severe shocks; it is identical in kind with that of the galvanic battery, the electricity of the under surface of the fish being the same with the negative pole, and that in the upper surface the same with the positive pole: its manner of action is, however, somewhat different, for, although the evolution of the electricity is continued for a sensible time, it is interrupted, being communicated by a succession of discharges.

SECTION XXX.

In order to explain the other methods of exciting electricity, and the recent discoveries that have been made in that science, it is necessary to be acquainted with the general theory of magnetism, and also with the magnetism of the earth, the director of the mariner's compass, and his guide through the ocean. Its influence extends over every part of the earth's surface, but its action on the magnetic needle determines the poles of this great magnet, which by no means coincide with the poles of the earth's rotation. In consequence of their attraction and repulsion, a needle freely suspended, whether it be magnetic or not, only remains in equilibrio when in the magnetic meridian, that is, in the plane which passes through the north and south magnetic poles. There are places where the magnetic meridian coincides with the terrestrial meridian; in these a magnetic needle freely suspended points to the true north; but if it be carried successively to different places on the earth's surface, its direction will deviate sometimes to the east and sometimes to the west of north. Lines drawn on the globe, through all the places where the needle points due north and south, are called lines of no variation, and they are extremely complicated. The direction of the needle is not even constant in the same place, but changes in a few years according to a law not yet determined. In 1657, the line of no variation passed through London; from that time it has moved slowly, but irregularly, westward, and is now in North America. In the year 1819, Sir Edward Parry, in his voyage to discover the north-west passage round America, sailed near the magnetic pole; and in 1824, Captain Lyon, on an expedition for the same purpose, found that the magnetic pole was then situate in 63° 26′ 51″ north latitude, and in 80° 51′ 25″ west longitude. It appears, from later researches, that the law of terrestrial magnetism is of considerable complexity and the existence of more than one magnetic pole in either hemisphere has been rendered highly probable; that there is one in Siberia seems to be decided by the recent observations of M. Hansteen,—it is in longitude 102° east of Greenwich, and a little to the north of the 60th degree of latitude: so that, by these data, the two magnetic poles in the northern hemisphere are about 180° distant from each other; but Captain Ross, who is just returned from a voyage in the polar seas, has ascertained that the American magnetic pole is in 70° 14′ north latitude, and 96° 40′ west longitude. The magnetic equator does not exactly coincide with the terrestrial equator; it appears to be an irregular curve inclined to the earth's equator at an angle of about 12°, and crossing it in at least three points in longitude 113° 14′ west, and 66° 46′ east of the meridian of Greenwich, and again somewhere between 156° 30′ of west longitude, and 116° east.

The needle is also subject to diurnal variations; in our latitudes it moves slowly eastward during the forenoon, and returns to its mean position about ten in the evening; it then deviates to the westward, and again returns to its mean position about ten in the morning. M. Kupffer, of Casan, ascertained, in the year 1831, that there is a nightly, as well as a diurnal variation, depending, in his opinion, upon a variation in the magnetic equator.

A magnetic needle, suspended so as to be moveable only in the vertical plane, dips, or becomes more and more inclined to the horizon the nearer it is brought to the magnetic pole, and there becomes vertical. At the magnetic equator it is horizontal, and between these two positions it assumes every degree of inclination. Captain

Lyon found that the dip in the latitude and longitude mentioned, very near the magnetic pole, was 86° 32′, and Captain Segelke determined it to be 69° 38′ at Woolwich in 1830. According to Captain Sabine, it appears to have been decreasing for the last fifty years at the rate of three minutes annually.

If a magnetised needle freely suspended, and at rest in the magnetic meridian, be drawn any number of degrees from its position, it will make a certain number of oscillations before it resumes its state of rest. The intensity of the magnetic force is determined from these oscillations in the same manner that the intensity of the gravitating and electrical forces are known from the vibrations of the pendulum and the balance of torsion, and in all these cases it is proportional to the square of the number of oscillations performed in a given time; consequently a comparison of the number of vibrations accomplished by the same needle, during the same time, in different parts of the earth's surface, will determine the variations in the magnetic action. By this method Humboldt and Rossel have discovered that the intensity of the magnetic force increases from the equator to the poles, where it is probably at its maximum. It appears to be doubled in the ascent from the equator to the western limits of Baffin's Bay. According to the magnetic observations of Professor Hansteen, of Christiania, the magnetic intensity has been decreasing annually at Christiania, London, and Paris, at the rate of its 235th, 725th, and 1020th parts, respectively, which he attributes to the revolution of the Siberian magnetic pole. There is, however, so much uncertainty in the magnetic phenomena of the earth, that the results require to be continually corrected by new observations.

The inventor of the mariner's compass, like most of the early benefactors of mankind, is unknown; it is even doubted which nation first made use of magnetic polarity to determine positions on the surface of the globe; but it is said that a rude form of the compass was invented in Upper Asia, and conveyed thence by the Tartars to China, where the Jesuit missionaries found traces of this instrument having been employed as a guide to land travellers in very remote antiquity. From that the compass spread over the east, and was imported into Europe by the Crusaders, and its construction improved by an artist of Amalfi, on the coast of Calabria. It seems that the Romans and Chinese only employed eight cardinal divisions, which the Germans successively bisected till there were thirty-two, and gave the points the names which they still bear.

The variation of the compass was unknown till Columbus, during his first voyage, observed that the needle declined from the meridian as he advanced across the Atlantic. The dip of the magnetic needle was first noticed by Robert Norman, in the year 1576.

Very delicate experiments have shown that all bodies are more or less susceptible of magnetism. Many of the gems give signs of it; cobalt, titanium, and nickel sometimes even possess the properties of attraction and repulsion; but the magnetic agency is most powerfully developed in iron, and in that particular ore of iron called the loadstone, which consists of the protoxide and the peroxide of iron, together with small portions of silica and alumina. A metal is often susceptible of magnetism if it only contains the 130000th part of its weight of iron, a quantity too small to be detected by any chemical test.

The bodies in question are naturally magnetic, but that property may be imparted by a variety of methods, as by friction with magnetic bodies, or juxtaposition to them, but none is more simple than percussion. A bar of hard steel, held in the direction of the dip, will become a magnet on receiving a few smart blows with a hammer on its upper extremity; and M. Hansteen has ascertained that every substance has magnetic poles when held in that position, whatever the materials may be of which it is composed.

One of the most distinguishing marks of magnetism is polarity, or the property a magnet possesses, when freely suspended, of spontaneously pointing nearly north and south, and always returning to that position when disturbed. Another property of a magnet is the attraction of unmagnetised iron. Both poles of a magnet attract iron, which in return attracts either pole of the magnet with an equal and contrary force. The magnetic intensity is most powerful at the poles, as may easily be seen by dipping the magnet into iron filings, which will adhere abundantly to each pole, while scarcely any attach themselves to the intermediate parts. The action of the magnet on unmagnetised iron is confined to attraction, whereas the reciprocal agency of magnets is characterized by a repulsive as well as an attractive force, for a north pole repels a north pole, and a south repels a south pole; but a north and a south pole mutually attract one another, which proves that there are two distinct kinds of magnetic forces, directly opposite in their effects, though similar in their mode of action.

Induction is the power which a magnet possesses of exciting temporary or permanent magnetism in such bodies in its vicinity as are capable of receiving it. By this property the mere approach of a magnet renders iron or steel magnetic, the more powerfully the less the distance. When the north pole of a magnet is brought near to, and in the line with an unmagnetised iron bar, the bar acquires all the properties of a perfect magnet, the end next the north pole of the magnet becomes a south pole, while the remote end becomes a north pole. Exactly the reverse takes place when the south pole is presented to the bar; so that each pole of a magnet induces the opposite polarity in the adjacent end of the bar, and the same polarity in the remote extremity; consequently the nearest extremity of the bar is attracted, and the farther repelled, but as the action is greater on the adjacent than on the distant part, the resulting force is that of attraction. By induction, the iron bar not only acquires polarity, but the power of inducing magnetism in a third body; and although all these properties vanish from the iron as soon as the magnet is removed, a lasting increase of intensity is generally imparted to the magnet itself by the reaction of the temporary magnetism of the iron. Iron acquires magnetism more rapidly than steel, yet it loses it as quickly on the removal of the magnet, whereas the steel is impressed with a lasting polarity.

A certain time is requisite for the induction of magnetism, and it may be accelerated by anything that excites a vibratory motion in the particles of the steel, such as the smart stroke of a hammer, or heat succeeded by sudden cold. A steel bar may be converted into a magnet by the transmission of an electric discharge through it, and as its efficacy is the same in whatever direction the electricity passes, the magnetism arises from its mechanical operation exciting a vibration among the particles of the steel. It has been observed that the particles of iron easily resume their neutral state after induction, but that those of steel resist the restoration of magnetic equilibrium, or a return to the neutral state: it is therefore evident, that any cause which removes or diminishes the resistance of the particles will tend to destroy the magnetism of the steel; consequently, the same mechanical means which develope magnetism will also destroy it. On that account, a steel bar may lose its magnetism by any mechanical concussion, such as by falling on a hard substance, a blow with a hammer, and heating to redness, which reduces the steel to the state of soft iron. The circumstances which determine whether it shall gain or lose being its position with respect to the magnetic equator, and the higher or lower intensity of its previous magnetic state.

Polarity of one kind only cannot exist in any portion of iron or steel, for in whatever manner the intensities of the two kinds of polarity may be diffused through a magnet, they exactly balance or compensate one another. The northern polarity is confined to one half of a magnet, and the southern to the other, and they are generally concentrated in or near the extremities of the bar. When a magnet is broken across its middle, each fragment is at once converted into a perfect magnet; the part which originally had a north pole, acquires a south pole at the fractured end, the part that originally had a south pole gets a north pole; and as far as mechanical division can be carried, it is found that each fragment, however small, is a perfect magnet.

A comparison of the number of vibrations accomplished by the same needle, during the same time, at different distances from a magnet, gives the law of magnetic intensity, which, like every known force that emanates from a centre, follows the inverse ratio of the square of the distance, a law that is not affected by the intervention of any substance whatever between the magnet and the needle, provided that substance be not itself susceptible of magnetism. Induction and the reciprocal action of magnets are, therefore, subject to the laws of mechanics, but the composition and resolution of the forces are complicated, in consequence of four forces being constantly in activity, two in each magnet.

The phenomena of magnetism may be explained on the hypothesis of two extremely rare fluids pervading all the particles of iron, and incapable of leaving them. Whether the particles of these fluids are coincident with the molecules of the iron, or that they only fill the interstices between them, is unknown and immaterial; but it is certain that the sum of all the magnetic molecules, added to the sum of all the spaces between them, whether occupied by matter or not, must be equal to the whole volume of the magnetic body. When the two fluids in question are combined they are inert, so that the substances containing them show no signs of magnetism; but when separate they are active, the molecules of each of the fluids attracting those of the opposite kind, and repelling those of the same kind. The decomposition of the united fluids is accomplished by the inductive influence of either of the separate fluids; that is to say, a ferruginous body acquires polarity by the approach of either the south or north pole of a magnet. The electric fluids are confined to the surfaces of bodies, whereas the magnetic fluids pervade each molecule of the mass; besides, the electric fluid has a perpetual tendency to escape, and does escape, when not prevented by the coercive power of the surrounding air and other non-conducting bodies. Such a tendency does not exist in the magnetic fluids, which never quit the substance that contains them under any circumstances whatever; nor is any sensible quantity of either kind of polarity ever transferred from one part to another of the same piece of steel. It appears that the two magnetic fluids, when decomposed by the influence of magnetising forces, only undergo a displacement to an insensible degree within the body. The action of all the particles so displaced upon a particle of the magnetic fluid in any particular situation, compose a resultant force, the intensity and direction of which it is the province of the analyst to determine. In this manner M. Poisson has proved that the result of the action of all the magnetic elements of a magnetised body is a force equivalent to the action of a very thin stratum covering the whole surface of a body, and consisting of the two fluids—the austral and the boreal, occupying different parts of it; or, in other words, the attractions and repulsions externally exerted by a magnet are exactly the same as if they proceeded from a very thin stratum of each fluid occupying the surface only, both fluids being in equal quantities, and so distributed that their total action upon all the points in the interior of the body are equal to nothing. Since the resulting force is the difference of the two polarities, its intensity must be greatly inferior to that of either.

It may be observed that, in addition to the forces already mentioned, there must be some coercive force analogous to friction which arrests the particles of both fluids, so as first to oppose the separation of the fluids, and then to prevent their reuniting. In soft iron the coercive force is either wanting or extremely feeble, since the iron is easily rendered magnetic by induction, and as easily loses its magnetism; whereas in steel the coercive force is extremely energetic, because it prevents the steel from acquiring the magnetic properties rapidly, and entirely hinders it from losing them when acquired. The feebleness of the coercive force in iron, and its energy in steel, with regard to the magnetic fluids, is perfectly analogous to the facility of transmission afforded to the electric fluids by non-electrics, and the resistance they experience in electrics. At every step the analogy between magnetism and electricity becomes more striking. The agency of attraction and repulsion is common to both, the positive and negative electricities are similar to the northern and southern polarities, and are governed by the same laws, namely, that between like powers there is repulsion, and between unlike powers there is attraction; each of these four forces is capable of acting most energetically when alone, but the electric equilibrium is restored by the union of the two electricities, and magnetic neutrality by the combination of the two polarities, thus respectively neutralizing each other when joined. All these forces vary inversely as the square of the distance, and consequently come under the same mechanical laws. A like analogy extends to magnetic and electrical induction. Iron and steel are in a state of equilibrium when the two magnetic polarities conceived to reside in them are equally diffused throughout the whole mass, so that they are altogether neutral. But this equilibrium is immediately disturbed on the approach of the pole of a magnet, which by induction transfers one kind of polarity to one end of the iron or steel bar, and the opposite kind to the other,—effects exactly similar to electrical induction. There is even a correspondence between the fracture of a magnet and that of an electric conductor; for if an oblong conductor be electrified by induction, its two extremities will have opposite electricities; and if in that state it be divided across the middle, the two portions, when removed to a distance from one another, will each retain the electricity that has been induced upon it. The analogy, however, does not extend to transference. A body may transfer a redundant quantity of positive or negative electricity to another, the one gaining at the expense of the other; but there is no instance of a body possessing only one kind of polarity. With this exception, there is such perfect correspondence between the theories of magnetic attractions and repulsions and electric forces in conducting bodies, that they not only are the same in principle, but are determined by the same formulæ. Experiment concurs with theory in proving the identity of these two unseen influences.

SECTION XXXI.

The disturbing effects of the aurora borealis and of lightning on the mariner's compass had been long known, but in the year 1819, M. Oersted, Professor of Natural Philosophy at Copenhagen, discovered that a current of voltaic electricity exerts a powerful influence on a magnetised needle, an observation which has given rise to the theory of electro-magnetism, the most interesting science of modern times, whether it be considered as leading us a step farther in generalization, by identifying two agencies hitherto referred to different causes, or as developing a new force unparalleled in the system of the world, which, overcoming the retardation from friction, and the obstacle of a resisting medium, maintains a perpetual motion, often vainly attempted, but which it seems altogether impossible to accomplish by means of any other force or combination of forces than the one in question.

When the two poles of a voltaic battery are connected by a metallic wire, so as to complete the circuit, the electricity flows without ceasing; and if a straight portion of that wire be placed parallel to, and horizontally above, a magnetised needle at rest in the magnetic meridian, but freely poised like the mariner's compass, the action of the electric current flowing through the wire will instantly cause the needle to change its position: its extremity will deviate from the north towards the east or west, according to the direction in which the current is flowing; and on reversing the direction of the current, the motion of the needle will be reversed also. The numerous experiments that have been made on the magnetic and electric fluids, as well as those on the various relative motions of a magnetic needle under the influence of galvanic electricity, arising from all possible positions of the conducting wire, and every direction of the voltaic current, together with all the other phenomena of electro-magnetism, are explained by Dr. Roget in some excellent articles on these subjects in the Library of Useful Knowledge.

All the experiments tend to prove that the force

emanating from the electric current, which produces such effects on the magnetic needle, acts at right angles to the current, and is therefore unlike any force hitherto known. The action of all the forces in nature is directed in straight lines, as far as we know, for the curves described by the heavenly bodies result from the composition of two forces, whereas, that which is exerted by an electrical current upon either pole of a magnet has no tendency to cause the pole to approach or recede, but to rotate about it. If the stream of electricity be supposed to pass through the centre of a circle whose plane is perpendicular to the current, the direction of the force exerted by the electricity will always be in the tangent to the circle, or at right angles to its radius; consequently the tangential force of the electricity has a tendency to make the pole of a magnet move in a circle round the wire of the battery. Mr. Barlow has proved that the action of each particle of the electric fluid in the wire, on each particle of the magnetic fluid in the needle, varies inversely as the square of the distance.

Rotatory motion was suggested by Dr. Wollaston; Mr. Faraday was the first who actually succeeded in making the pole of a magnet rotate about a vertical conducting wire. In order to limit the action of the electricity to one pole, about two-thirds of a small magnet was immersed in mercury, the lower end being fastened by a thread to the bottom of the vessel containing the mercury. When the magnet was thus floating almost vertically with its north pole above the surface, a current of positive electricity was made to descend perpendicularly through a wire touching the mercury, and immediately the magnet began to rotate from left to right about the wire. As the force is uniform, the rotation was accelerated till the tangential force was balanced by the resistance of the mercury, when it became constant. Under the same circumstances, the south pole of the magnet rotates from right to left. It is evident from this experiment that the wire may also be made to perform a rotation round the magnet, since the action of the current of electricity on the pole of the magnet must necessarily be accompanied by a corresponding reaction of the pole of the magnet on the electricity in the wire. This experiment has been accomplished by a vast number of contrivances, and even a small battery, consisting of two plates, has performed the rotation. Mr. Faraday produced both motions at the same time in a vessel containing mercury; the wire and the magnet revolved in one direction about a common centre of motion, each following the other.

The next step was to make a magnet and

also a cylinder revolve about their own axes, which they do with great rapidity. Mercury has been made to rotate by means of voltaic electricity, and Professor Ritchie has exhibited in the Royal Institution the singular spectacle of the rotation of water by the same means, while the vessel containing it remained stationary. The water was in a hollow double cylinder of glass, and on being made the conductor of electricity, was observed to revolve in a regular vortex, changing its direction as the poles of the battery were alternately reversed. Professor Ritchie found that all the different conductors hitherto tried by him, such as water, charcoal, &c. give the same electro-magnetic results, when transmitting the same quantity of electricity, and that they deflect the magnetic needle in an equal degree when their respective axes of conduction are at the same distance from it. But one of the most extraordinary effects of the new force is exhibited by coiling a copper wire, so as to form a helix or corkscrew, and connecting the extremities of the wires with the poles of a galvanic battery. If a magnetised steel bar or needle be placed within the screw, so as to rest upon the lower and interior part, the instant a current of electricity is sent through the wire of the helix, the steel bar starts up by the influence of this invisible power, and remains suspended in the air in opposition to the force of gravitation. The effect of the electro-magnetic power exerted by each turn of the wire is to urge the north pole of the magnet in one direction, and the south pole in the other; the force thus exerted is multiplied in degree and increased in extent by each repetition of the turns of the wire, and in consequence of these opposing forces the bar remains suspended. This helix has all the properties of a magnet while the electrical current is flowing through it, and may be substituted for one in almost every experiment. It acts as if it had a north pole at one extremity and a south pole at the other, and is attracted and repelled by the poles of a magnet exactly as if it were one itself. All these effects depend upon the course of the electricity, that is, on the direction of the turns of the screw, according as they are from right to left, or from left to right, being in the one case exactly the contrary of what it is in the other.

The effects of electricity in motion on magnets are not only precisely the same as the reciprocal action of magnetised bodies, but its influence in inducing magnetism in unmagnetised iron and steel is also the same with magnetic induction. The term induction, when applied to electric currents, expresses the power which these currents possess of inducing any particular state upon matter in their immediate neighbourhood, otherwise neutral or indifferent. For example, the connecting wire of a galvanic battery holds iron filings suspended like an artificial magnet, as long as the current continues to flow through it; and the most powerful temporary magnets that have ever been made are obtained by bending a thick cylinder of soft iron into the form of a horseshoe, and surrounding it with a coil of thick copper wire covered with silk, to prevent communication between its parts. When this wire forms part of a galvanic circuit, the iron becomes so highly magnetic, that a temporary magnet of this kind made by Professor Henry of the Albany Academy, in the United States, sustained nearly a ton weight. The iron loses its magnetic power the instant the electricity ceases to circulate, and acquires it again as instantaneously when the circuit is renewed. Steel needles are rendered permanently magnetic by electrical induction; the effect is produced in a moment, and as readily by juxtaposition as by contact; the nature of the poles depends upon the direction of the current, and the intensity is proportional to the quantity of electricity.

It appears from what precedes, that the principle and characteristic phenomena of the electro-magnetic science are, the evolution of a tangential and rotatory force exerted between a conducting body and a magnet; and the transverse induction of magnetism by the conducting body in such substances as are susceptible of it.

The action of an electric current causes a deviation of the compass from the plane of the magnetic meridian. In proportion as the needle recedes from the meridian, the intensity of the force of terrestial magnetism increases, while at the same time the electro-magnetic force diminishes; the number of degrees at which the needle stops, and which mark where the equilibrium between these two forces takes place, will indicate the intensity of the galvanic current. The galvanometer, constructed upon this principle, is employed to measure the intensity of galvanic currents collected and conveyed to it by wires. This instrument is rendered much more sensible by neutralizing the effects of the earth's magnetism on the needle, which is accomplished by placing a second magnetised needle so as to counteract the action of the earth on the first, a precaution requisite in all delicate magnetical experiments.

SECTION XXXII.

The science of electro-magnetism which has been under consideration, and must render the name of M. Oersted ever memorable, relates to the reciprocal action of electrical and magnetic currents: M. Ampère, by discovering the mutual action of electrical currents on one another, has added a new branch to the subject, to which he has given the name of electro-dynamics.

When electric currents are passing through two conducting wires so suspended or supported as to be capable of moving both towards and from one another, they show mutual attraction or repulsion, according as the currents are flowing in the same or in contrary directions; the phenomena varying with the relative inclinations and positions of the streams of electricity. It appears that the mutual action of such currents, whether they flow in the same or in contrary directions, whether they be parallel, perpendicular, diverging, converging, circular, or heliacal, all produce different kinds of motion, in a conducting wire, both rectilineal and circular, and also the rotation of a wire helix, such as that described and now called an electro-dynamic cylinder on account of some improvements in its construction; and as the hypothesis of a force varying inversely as the square of the distance accords perfectly with all the observed phenomena, these motions come under the same laws of dynamics and analysis as any other branch of physics.

The theory of electro-dynamics, as well as actual experiment, confirms the identity between the agencies of electro-dynamic cylinders, or helices, and magnets. The law of the reciprocal action of a cylinder and an electric current is precisely the same, and all the experiments that can be performed with the cylinder might be accomplished with a magnet. It has already been observed that the two extremities of an electro-dynamic cylinder or helix exhibit all the properties possessed by the poles of a magnet; that end in which the current of positive electricity is moving in a direction similar to the motion of the hands of a watch, acting as a south pole, and the other end, in which the current is flowing in a contrary direction, exhibiting northern polarity. In conformity with this resemblance, electro-dynamic cylinders act on each other precisely as if they were magnets, during the time the electricity is flowing through them.

The phenomena mark a very decided difference between the action of electricity in motion or at rest, that is, between voltaic and common electricity; the laws they follow are in many respects of an entirely different nature. Since voltaic electricity flows perpetually, it cannot be accumulated, and consequently has no tension or tendency to escape from the wires which conduct it. Nor do these wires either attract or repel light bodies in their vicinity, whereas ordinary electricity can be accumulated in insulated bodies to a great degree, and in that state of rest the tendency to escape is proportional to the quantity accumulated and the resistance it meets with. In ordinary electricity, the law of action is, that dissimilar electricities attract, and similar electricities repel one another. In voltaic electricity, on the contrary, similar currents, or such as are moving in the same direction, attract one another, while a mutual repulsion is exerted between dissimilar currents, or such as flow in opposite directions. The common electricity escapes when the pressure is removed, but the electro-dynamical effects are the same whether the conductors be in air or in vacuo.

Although the effects produced by a current of electricity depend upon the celerity of its motion, the velocity with which it moves through a conducting wire is unknown. We are equally ignorant whether it be uniform or varied, but the method of transmission has a marked influence on the results; for when it flows without intermission, it occasions a deviation in the magnetic needle, but it has no effect whatever when its motion is discontinuous or interrupted, like the current produced by the common electrical machine when a communication is made between the positive and negative conductors.

M. Ampère has established a theory of electro-magnetism suggested by the analogy between electro-dynamic cylinders and magnets, founded upon the reciprocal attraction of electric currents, to which all the phenomena of magnetism and electro-magnetism may be reduced, by assuming that the magnetic properties which bodies possess derive these properties from currents of electricity circulating about every part in one uniform direction. It has been observed that, although every particle of a magnet possess like properties with the whole, yet the general effect is the same as if the magnetic properties were confined to the surface: consequently the internal electro-currents must compensate one another, and therefore the magnetism of a body is supposed to arise from a superficial current of electricity constantly circulating in a direction perpendicular to the axis of the magnet; so that the reciprocal action of magnets, and all the phenomena of electro-magnetism, are reduced to the action and reaction of superficial currents of electricity acting at right angles to the direction of the currents. Notwithstanding some experiments made by M. Ampère to elucidate the subject, there is still an uncertainty in the theory of the induction of magnetism by an electric current in a body near it; for it does not appear whether electric currents which did not previously exist are actually produced by induction, or if its effect be only to give one uniform direction to the infinite number of electric currents previously existing in the particles of the body, and thus rendering them capable of exhibiting magnetic phenomena, in the same manner as polarization reduces those undulations of light to one plane which had previously been performed in every plane. Possibly both may be combined in producing the effect; for the action of an electric current may not only give a common direction to those already existing, but may also increase their intensity. However that may be, by assuming that the attraction and repulsion of the elementary portions of electric currents vary inversely as the square of the distance, the action being at right angles to the direction of the current, it is found that the attraction and repulsion of a current of indefinite length on the elementary portion of a parallel current at any distance from it, is in the simple ratio of the shortest distance between them; consequently the reciprocal action of electric currents is reduced to the composition and resolution of forces, so that the phenomena of electro-magnetism are brought under the laws of dynamics by the theory of Ampère.

SECTION XXXIII.

From the law of action and reaction being equal and contrary, it might be expected that, as electricity powerfully affects magnets, so, conversely, magnetism ought to produce electrical phenomena. By proving this very important fact from a series of highly interesting and ingenious experiments, Mr. Faraday has added another branch to the science, which he has named magneto-electricity. A great quantity of copper wire was coiled in the form of a helix round one half of a ring of soft iron, and connected with a galvanic battery, while a similar helix connected with a galvanometer was wound round the other half of the ring, but not touching the first helix. As soon as contact was made with the battery, the needle of the galvanometer was deflected, but the action was transitory, for when the contact was continued the needle returned to its usual position, and was not affected by the continual flow of the electricity through the wire connected with the battery. As soon, however, as the contact was broken, the needle of the galvanometer was again deflected, but in the contrary direction. Similar effects were produced by an apparatus consisting of two helices of copper wire coiled round a block of wood, instead of iron, from which Mr. Faraday infers that the electric current passing from the battery through one wire induces a similar current through the other wire, but only at the instant of contact, and that a momentary current is induced in a contrary direction when the passage of the electricity is suddenly interrupted. These brief currents or waves of electricity were found to be capable of magnetizing needles, of passing through a small extent of fluid, and when charcoal points were interposed in the current of the induced helix, a minute spark was perceived as often as the contacts were made or broken, but neither chemical action nor any other electric effects were obtained. A deviation of the needle of the galvanometer took place when common magnets were employed instead of the voltaic current; so that the magnetic and electric fluids are identical in their effects in this interesting experiment. Again, when a helix formed of 220 feet of copper wire, into which a cylinder of soft iron was introduced, was placed between the north and south poles of two bar magnets, and connected with the galvanometer by means of wires from each extremity, as often as the magnets were brought into contact with the iron cylinder, it became magnetic by induction, and produced a deflection in the needle of the galvanometer. On continuing the contact, the needle resumed its natural position, and when the contact was broken, the deflection took place in the opposite direction; when the magnetic contacts were reversed, the deflection was reversed also. With strong magnets, so powerful was the action, that the needle of the galvanometer whirled round several times successively; and similar effects were produced by the mere approximation or removal of the helix to the poles of the magnets. Thus magnets produce the very same effects on the galvanometer that electricity does. Though at that time no chemical decomposition was effected by these momentary currents which emanated from the magnets, they agitated the limbs of a frog, and Mr. Faraday justly observes, that 'an agent which is conducted along metallic wires in the manner described, which, whilst so passing, possesses the peculiar magnetic actions and force of a current of electricity, which can agitate and convulse the limbs of a frog, and which finally can produce a spark by its discharge through charcoal, can only be electricity.' Hence it appears that electrical currents are evolved by magnets, which produce the same phenomena with the electrical currents from the voltaic battery; they, however, differ materially in this respect—that time is required for the exercise of the magneto-electric induction, whereas volta-electric induction is instantaneous.

After Mr. Faraday had proved the identity of the magnetic and electric fluids by producing the spark, heating metallic wires, and accomplishing chemical decomposition, it was easy to increase these effects by more powerful magnets and other arrangements. The following apparatus is now in use, which is in effect a battery, where the agent is the magnetic, instead of the voltaic fluid, or, in other words, electricity.

A very powerful horse-shoe magnet, formed of twelve steel plates in close approximation, is placed in a horizontal position. An armature consisting of a bar of the purest soft iron has each of its ends bent at right angles, so that the faces of those ends may be brought directly opposite and close to the poles of the magnet when required. Two series of copper wires—covered with silk, in order to insulate them—are wound round the bar of soft iron as compound helices. The extremities of these wires, having the same direction, are in metallic connexion with a circular disc, which dips into a cup of mercury, while the ends of the wires in the opposite direction are soldered to a projecting screw-piece, which carries a slip of copper with two opposite points. The steel magnet is stationary; but when the armature, together with its appendages, is made to rotate horizontally, the edge of the disc always remains immersed in the mercury, while the points of the copper slip alternately dip in it and rise above it. By the ordinary laws of induction, the armature becomes a temporary magnet while its bent ends are opposite the poles of the steel magnet, and ceases to be magnetic when they are at right angles to them. It imparts its temporary magnetism to the helices which concentrate it; and while one set conveys a current to the disc, the other set conducts the opposite current to the copper slip. But as the edge of the revolving disc is always immersed in the mercury, one set of wires is constantly maintained in contact with it, and the circuit is only completed when a point of the copper slip dips in the mercury also; but the circuit is broken the moment that point rises above it. Thus, by the rotation of the armature, the circuit is alternately broken and renewed; and as it is only at these moments that electric action is manifested, a brilliant spark takes place every time the copper point touches the surface of the mercury. Platina wire is ignited, shocks smart enough to be disagreeable are given, and water is decomposed with astonishing rapidity, by the same means, which proves beyond a doubt the identity of the magnetic and electric agencies, and places Mr. Faraday, whose experiments established the principle, in the first rank of experimental philosophers.

SECTION XXXIV.

M. Arago discovered an entirely new source of magnetism in rotatory motion. If a circular plate of copper be made to revolve immediately above or below a magnetic needle or magnet, suspended in such a manner that the needle may rotate in a plane parallel to that of the copper plate, the magnet tends to follow the circumvolution of the plate; or if the magnet revolves, the plate tends to follow its motion; and so powerful is the effect, that magnets and plates of many pounds weight have been carried round. This is quite independent of the motion of the air, since it is the same when a pane of glass is interposed between the magnet and the copper. When the magnet and the plate are at rest, not the smallest effect, attractive, repulsive, or of any kind, can be perceived between them. In describing this phenomenon, M. Arago states that it takes place not only with metals, but with all substances, solids, liquids, and even gases, although the intensity depends upon the kind of substancc in motion. Experiments recently made by Mr. Faraday explain this singular action. A plate of copper, twelve inches in diameter and one-fifth of an inch thick, was placed between the poles of a powerful horse-shoe magnet, and connected at certain points with a galvanometer by copper wires. When the plate was at rest no effect was produced, but as soon as the plate was made to revolve rapidly, the galvanometer needle was deflected sometimes as much as 90°, and by a uniform rotation, the deflection was constantly maintained at 45°. When the motion of the copper plate was reversed, the needle was deflected in the contrary direction, and thus a permanent current of electricity was evolved by an ordinary magnet. The intensity of the electricity collected by the wires, and conveyed by them to the galvanometer, varied with the position of the plate relatively to the poles of the magnet.

The motion of the electricity in the copper plate may be conceived, by considering, that merely from moving a single wire like the spoke of a wheel before a magnetic pole, a current of electricity tends to flow through it from one end to the other; hence, if a wheel be constructed of a great many such spokes, and revolved near the pole of a magnet in the manner of the copper disc, each radius or spoke will tend to have a current produced in it as it passes the pole. Now, as the circular plate is nothing more than an infinite number of radii or spokes in contact, the currents will flow in the direction of the radii if a channel be open for their return, and in a continuous plate that channel is afforded by the lateral portions on each side of the particular radius close to the magnetic pole. This hypothesis is confirmed by observation, for the currents of positive electricity set from the centre to the circumference, and the negative from the circumference to the centre, and vice versâ, according to the position of the magnetic poles and the direction of rotation. So that a collecting wire at the centre of the copper plate conveys positive electricity to the galvanometer in one case, and negative in another; that collected by a conducting wire in contact with the circumference of the plate is always the opposite of the electricity conveyed from the centre. It is evident that when the plate and magnet are both at rest, no effect takes place, since the electric currents which cause the deflection of the galvanometer cease altogether. The same phenomena may be produced by electro-magnets. The effects are the same when the magnet rotates and the plate remains at rest. When the magnet revolves uniformly about its own axis, electricity of the same kind is collected at its poles, and the opposite electricity at its equator.

The phenomena which take place in M. Arago's experiments may be explained on this principle, for when both the copper plate and the magnet are revolving, the action of the electric current, induced in the plate by the magnet in consequence of their relative motion, tends continually to diminish that relative motion; that is, to bring the moving bodies into a state of relative rest, so that if one be made to revolve by an extraneous force, the other will tend to revolve about it in the same direction, and with the same velocity.

When a plate of iron, or of any substance capable of being made either a temporary or permanent magnet, revolves between the poles of a magnet, it is found that dissimilar poles on opposite sides of the plate neutralize each other's effects, so that no electricity is evolved, while similar poles on each side of the revolving plate increase the quantity of electricity, and a single pole end-on is sufficient. But when copper, and substances not sensible to ordinary magnetic impressions, revolve, similar poles on opposite sides of the plate neutralize each other, dissimilar poles on each side exalt the action, and a single pole at the edge of the revolving plate, or end-on, does nothing. This forms a test for distinguishing the ordinary magnetic force from that produced by rotation. If unlike poles, that is a north and a south pole, produce more effect than one pole, the force will be due to electric currents; if similar poles produce more effect than one, then the power is not electric. These investigations show that there are really very few bodies magnetic in the manner of iron. Mr. Faraday therefore arranges substances in three classes, with regard to their relation to magnets. Those affected by the magnet when at rest like iron, steel, and nickel, which possess ordinary magnetic properties; those affected when in motion, in which electric currents are evolved by the inductive force of the magnet, such as copper; and lastly, those which are perfectly indifferent to the magnet, whether at rest or in motion.

It has already been observed, that three bodies are requisite to form a galvanic circuit, one of which must be fluid; but in 1822, Professor Seebeck, of Berlin, discovered that electric currents may be produced by the partial application of heat to a circuit formed of two solid conductors. For example, when a semicircle of bismuth, joined to a semicircle of antimony, so as to form a ring, is heated at one of the junctions by a lamp, a current of electricity flows through the circuit from the antimony to the bismuth, and such thermo-electric currents produce all the electro-magnetic effects. A compass needle placed either within or without the circuit, and at a small distance from it, is deflected from its natural position, in a direction corresponding to the way in which the electricity is flowing. If such a ring be suspended so as to move easily in any direction, it will obey the action of a magnet brought near it, and may even be made to revolve. According to the researches of M. Nobili, the same substance, unequally heated, exhibits electrical currents. The experiments of Professor Cumming show that the mutual action of a magnet and a thermo-electric current, is subject to the same laws as those of magnets and galvanic currents, consequently all the phenomena of repulsion, attraction, and rotation may be exhibited by a thermo-electric current. It is, however, so feeble, that neither heat, the spark, nor chemical action have been observed, nor can repulsion, attraction of light substances at sensible distances, or any other effects of tension, be perceived.

SECTION XXXV.

In all the experiments hitherto described, artificial magnets alone were used, but it is obvious that the magnetism of the terrestrial spheroid which has so powerful an influence on the mariner's compass, must also affect electrical currents. It consequently appears that a piece of copper wire bent into a rectangle, and free to revolve on a vertical axis, arranges itself with its plane at right angles to the magnetic meridian, as soon as a stream of electricity is sent through it. Under the same circumstances a similar rectangle, suspended on a horizontal axis at right angles to the magnetic meridian, assumes the same inclination with the dipping needle. So that terrèstrial magnetism has the same influence on electrical currents as an artificial magnet. But the magnetic action of the earth also induces electric currents. When a hollow helix of copper wire, whose extremities are connected with the galvanometer, is placed in the magnetic dip, and suddenly inverted several times, accommodating the motion to the oscillations of the needle, the latter is soon made to vibrate through an arc of 80° or 90°. Hence it is evident, that whatever may be the cause of terrestrial magnetism, it produces currents of electricity by its direct inductive power upon a metal not capable of exhibiting any of the ordinary magnetic properties. The action on the galvanometer is much greater when a cylinder of soft iron is inserted into the helix, and the same results follow the simple introduction of the iron cylinder into, or removal out of the helix. These effects arise from the iron being made a temporary magnet by the inductive action of terrestrial magnetism, for a piece of iron, such as a poker, becomes a magnet for the time, when placed in the line of the magnetic dip.

M. Biot has formed a theory of terrestrial magnetism upon the observations of M. de Humboldt as data. Assuming that the action of the two opposite magnetic poles of the earth upon any point is inversely as the square of the distance, he obtains a general expression for the direction of the magnetic needle, depending upon the distance between the north and south magnetic poles; so that if one of these quantities varies, the corresponding variation of the other will be known. By making the distance between the poles vary, and comparing the resulting direction of the needle with the observations of M. de Humboldt, he found that the nearer the poles are supposed to approach to one another, the more did the computed and observed results agree ; and when the poles were assumed to coincide, or nearly so, the difference between theory and observation was the least possible. It is evident, therefore, that the earth does not act as if it were a permanently magnetic body, the distinguishing characteristic of which is, to have two poles at a distance from one another. Mr. Barlow has investigated this subject with much skill and success. He first proved that the magnetic power of an iron sphere resides in its surface ; he then inquired what the superficial action of an iron sphere in a state of transient magnetic induction, on a magnetised needle, would be, if insulated from the influence of terrestrial magnetism. The results obtained, corroborated by the profound analysis of M. Poisson, on the hypothesis of the two poles being indefinitely near the centre of the sphere, are identical with those obtained by M. Biot for the earth from M. de Humboldt's observations. Whence it follows, that the laws of terrestrial magnetism deduced from the formulæ of M. Biot, are inconsistent with those which belong to a permanent magnet, but that they are perfectly concordant with those belonging to a body in a state of transient magnetic induction. It appears, therefore, that the earth is to be considered as only transiently magnetic by induction, and not a real magnet. Mr. Barlow has rendered this extremely probable by forming a wooden globe, with grooves admitting of a copper wire being coiled round it parallel to the equator from pole to pole. When a current of electricity was sent through the wire, a magnetic needle suspended above the globe, and neutralized from the influence of the earth's magnetism, exhibited all the phenomena of the dipping and variation needles, according to its positions with regard to the wooden globe. As there can be no doubt that the same phenomena would be exhibited by currents of thermo, instead of voltaic, electricity, if the grooves of the wooden globe were filled by rings constituted of two metals, it seems highly probable that the heat of the sun may be the great agent in developing electric currents in or near the surface of the earth, by its action upon the substances of which the globe is composed, and, by the changes in its intensity, may occasion the diurnal variation of the compass, and the other vicissitudes in terrestrial magnetism evinced by the disturbance in the directions of the magnetic lines, in the same manner as it influences the parallelism of the isothermal lines. That such currents do exist in metalliferous veins appears from the experiments of Mr. Robert Fox in the Cornish copper-mines. However, it is probable that the secular and periodic disturbances in the magnetic force are occasioned by a variety of combining circumstances. Among others, M. Biot mentions the vicinity of mountain chains to the place of observation, and still more the action of extensive volcanic fires, which change the chemical state of the terrestrial surface, they themselves varying from age to age, some becoming extinct, while others burst into activity.

It is moreover probable that terrestrial magnetism may be owing, to a certain extent, to the earth's rotation. Mr. Faraday has proved that all the phenomena of revolving plates may be produced by the inductive action of the earth's magnetism alone. If a copper plate be connected with a galvanometer by two copper wires, one from the centre and another from the circumference, in order to collect and convey the electricity, it is found that, when the plate revolves in a plane passing through the line of the dip, the galvanometer is not affected; but as soon as the plate is inclined to that plane, electricity begins to be developed by its rotation; it becomes more powerful as the inclination increases, and arrives at a maximum when the plate revolves at right angles to the line of the dip. When the revolution is in the same direction with that of the hands of a watch, the current of electricity flows from its centre to the circumference; and when the rotation is in the opposite direction, the current sets the contrary way. The greatest deviation of the galvanometer amounted to 50° or 60°, when the direction of the rotation was accommodated to the oscillations of the needle. Thus a copper plate, revolving in a plane at right angles to the line of the dip, forms a new electrical machine, differing from the common plate-glass machine, by the material of which it is composed being the most perfect conductor, whereas glass is the most perfect non-conductor; besides, insulation, which is essential in the glass machine, is fatal in the copper one. The quantity of electricity evolved by the metal does not appear to be inferior to that developed by the glass, though very different in intensity.

From the experiments of Mr. Faraday, and also from theory, it is possible that the rotation of the earth may produce electric currents in its own mass. In that case, they would flow superficially in the meridians, and if collectors could be applied at the equator and poles, as in the revolving plate, negative electricity would be collected at the equator, and positive at the poles; but without something equivalent to conductors to complete the circuit, these currents could not exist.

Since the motion, not only of metals but even of fluids, when under the influence of powerful magnets, evolves electricity, it is probable that the gulf stream may exert a sensible influence upon the forms of the lines of magnetic variation, in consequence of electric currents moving across it, by the electro-magnetic induction of the earth. Even a ship passing over the surface of the water, in northern or southern latitudes, ought to have electric currents running directly across the line of her motion. Mr. Faraday observes, that such is the facility with which electricity is evolved by the earth's magnetism, that scarcely any piece of metal can be moved in contact with others without a development of it, and that consequently, among the arrangements of steam engines and metallic machinery, curious electro-magnetic combinations probably exist, which have never yet been noticed.

What magnetic properties the sun and planets may have, it is impossible to conjecture, although their rotation might lead us to infer that they are similar to the earth in this respect. According to the observations of MM. Biot and Gay-Lussac, during their aërostatic expedition, the magnetic action is not confined to the surface of the earth, but extends into space. A decrease in its intensity was perceptible, and as it most likely follows the ratio of the inverse square of the distance, it must extend indefinitely. It is probable that the moon has become highly magnetic by induction, in consequence of her proximity to the earth, and because her greatest diameter always points towards it. Should the magnetic, like the gravitating force, extend through space, the induction of the sun, moon, and planets must occasion perpetual variations in the intensity of terrestrial magnetism, by the continual changes in their relative positions.

In the brief sketch that has been given of the five kinds of electricity, those points of resemblance have been pointed out which are characteristic of one individual power; but as many anomalies have been lately removed, and the identity of the different kinds placed beyond a doubt, by Mr. Faraday, it may be satisfactory to take a summary view of the various coincidences in their modes of action on which their identity has been so ably and completely established by that great electrician.

The points of comparison are attraction and repulsion at sensible distances, discharge from points through air, the heating power, magnetic influence, chemical decomposition, action on the human frame, and lastly the spark.

Attraction and repulsion at sensible distances, which are so eminently characteristic of ordinary electricity, and, in a lesser degree, also, of the voltaic and magnetic currents, have not been perceived in either the thermo or animal electricities, not on account of difference of kind, but entirely owing to inferiority in tension; for even the ordinary electricity, when much reduced in quantity and intensity, is incapable of exhibiting these phenomena.

Ordinary electricity is readily discharged from points through air, but Mr. Faraday found that no sensible effect took place from a battery consisting of 140 double plates, either through air or in the exhausted receiver of an air-pump, the tests of the discharge being the electrometer and chemical action,—a circumstance entirely owing to the small degree of tension, for an enormous quantity of electricity is required to make these effects sensible, and for that reason they cannot be expected from the other kinds, which are much inferior in degree. Common electricity passes easily through rarefied and hot air, and also through flame. Mr. Faraday effected chemical decomposition and a deflection of the galvanometer by the transmission of voltaic electricity through heated air, and observes that these experiments are only cases of the discharge which takes place through air between the charcoal terminations of the poles of a powerful battery when they are gradually separated after contact—for the air is then heated; and Sir Humphry Davy mentions that, with the original voltaic apparatus at the Royal Institution, the discharge passed through four inches of air; that, in the exhausted receiver of an air-pump, the electricity would strike through nearly half an inch of space, and that the combined effects of rarefaction and heat were such, upon the included air, as to enable it to conduct the electricity through a space of six or seven inches. A Leyden jar may be instantaneously charged with voltaic, and also with magneto-electricity—another proof of their tension. Such effects cannot be obtained from the other kinds, on account of their weakness only.

The heating powers of ordinary and voltaic electricity have long been known, but the world is indebted to Mr. Faraday for the wonderful discovery of the heating power of the magnetic fluid: there is no indication of heat either from the animal or thermo-electricities. All the kinds of electricity have strong magnetic powers, those of the voltaic fluid are highly exalted, and the existence of the magneto and thermo-electricities was discovered by their magnetic influence alone. The needle has been deflected by all in the same manner, and, with the exception of thermo-electricity, magnets have been made by all according to the same laws. Ordinary electricity was long supposed incapable of deflecting the needle, and it required all Mr. Faraday's ingenuity to produce that effect. He has, however, proved that, in this respect, also, ordinary electricity agrees with voltaic, but that time must be allowed for its action. It deflected the needle, whether the current was sent through rarefied air, water, or wire. Numerous chemical decompositions have been effected by ordinary and voltaic electricity, according to the same laws and modes of arrangement. Dr. Davy decomposed water by the electricity of the torpedo,—Mr. Faraday accomplished its decomposition, and Dr.

Ritchie its composition, by means of magnetic action; but the chemical effects of the thermo-electricity have not yet been observed. The electric and galvanic shock, the flash in the eyes, and the sensation on the tongue, are well known. All these effects are produced by magneto-electricity, even to a painful degree. The torpedo and gymnotus electricus give severe shocks, and the limbs of a frog have been convulsed by thermo-electricity. The last point of comparison is the spark, which is already mentioned as common to the ordinary, voltaic, and magnetic fluids; and although it has not yet been seen from the thermo and animal electricities, there can be no doubt that it is only on account of their feebleness. Indeed, the conclusion drawn by Mr. Faraday is that the five kinds of electricity are identical, and that the differences of intensity and quantity are quite sufficient to account for what were supposed to be their distinctive qualities. He has given still greater assurance of their identity by showing that the magnetic force and the chemical action of electricity are in direct proportion to the absolute quantity of the fluid which passes through the galvanometer, whatever its intensity may be.

In light, heat, and electricity, or magnetism, nature has exhibited principles which do not occasion any appreciable change in the weight of bodies, although their presence is manifested by the most remarkable mechanical and chemical action. These agencies are so connected, that there is reason to believe they will ultimately be referred to some one power of a higher order, in conformity with the general economy of the system of the world, where the most varied and complicated effects are produced by a small number of universal laws. These principles penetrate matter in all directions; their velocity is prodigious, and their intensity varies inversely as the square of the distance. The development of electric currents, as well by magnetic as electric induction, the similarity in their mode of action in a great variety of circumstances, but above all the production of the spark from a magnet, the ignition of metallic wires, and chemical decomposition, show that magnetism can no longer be regarded as a separate, independent principle. That light is visible heat seems highly probable; and although the evolution of light and heat during the passage of the electric fluid may be from the compression of the air, yet the development of electricity by heat, the influence of heat on magnetic bodies, and that of light on the vibrations of the compass, show an occult connexion between all these agents, which probably will one day be revealed; and in the mean time it opens a noble field of experimental research to philosophers of the present, perhaps of future ages.

SECTION XXXVI.

In considering the constitution of the earth and the fluids which surround it, various subjects have presented themselves to our notice, of which some, for aught we know, are confined to the planet we inhabit; some are common to it and to the other bodies of our system; but an all-pervading ether probably fills the whole visible creation, and conveys, in the form of light, tremors which may have been excited in the deepest recesses of the universe thousands of years before we were called into being. The existence of such a medium, though at first hypothetical, is nearly proved by the undulatory theory of light, and rendered all but certain, within a few years, by the motion of comets, and by its action upon the vapours of which they are chiefly composed. It has often been imagined that, in addition to the effects of heat and electricity, the tails of comets have infused new substances into our atmosphere. Possibly the earth may attract some of that nebulous matter, since the vapours raised by the sun's heat, when the comets are in perihelio, and which form their tails, are scattered through space in their passage to their aphelion; but it has hitherto produced no effect, nor have the seasons ever been influenced by these bodies. In all probability, the tails of comets may have passed over the earth without its inhabitants being conscious of their presence.

The passage of comets has never sensibly disturbed the stability of the solar system; their nucleus, being in general only a mass of vapours, is so rare, and their transit so rapid, that the time has not been long enough to admit of a sufficient accumulation of impetus to produce a perceptible action. Indeed, M. Dusejour has proved that, under the most favourable circumstances, a comet cannot remain longer than two hours and a half at a less distance than 10500 leagues from the earth. The comet of 1770 passed within about six times the distance of the moon from the earth, without even affecting our tides; and as the moon has no sensible influence on the equilibrium of the atmosphere, a comet must have still less. According to La Place, the action of the earth on the comet of 1770 augmented the period of its revolution by more than two days; and if comets had any perceptible disturbing energy, the reaction of the comet ought to have increased the length of our year. Had the mass of that comet been equal to the mass of the earth, its disturbing action would have increased the length of the sideral year by 2^h 53^m; but as Delambre's computations from the Greenwich observations of the sun, show that the length of the year has not been increased by the fraction of a second, its mass could not have been equal to the $\frac{1}{5000}$ part of that of the earth. This accounts for the same comet having twice swept through the system of Jupiter's satellites without deranging the motions of these moons. Dusejour has computed that a comet, equal in mass to the earth, passing at the distance of 12150 leagues from our planet, would increase the length of the year to 367^d 16^h 5^m, and the obliquity of the ecliptic as much as 2°. So the principal action of comets would be to alter the calendar, even if they were dense enough to affect the earth.

Comets traverse all parts of the heavens; their paths have every possible inclination to the plane of the ecliptic, and, unlike the planets, the motion of more than half of those that have appeared have been retrograde. They are only visible when near their perihelia; then their velocity is such, that its square is twice as great as that of a body moving in a circle at the same distance, they consequently remain a very short time within the planetary orbits; and as all the conic sections of the same focal distance sensibly coincide, through a small arc on each side of the extremity of their axis, it is difficult to ascertain in which of these curves the comets move, from observations made, as they necessarily must be, at their perihelia; but probably they all move in extremely excentric ellipses, although in most cases the parabolic curve coincides most nearly with their observed motions. Some few seem to describe hyperbolas; such being once visible to us, would vanish for ever, to wander through boundless space, to the remote systems of the universe. If a planet be supposed to revolve in a circular orbit, whose radius is equal to the perihelion distance of a comet moving in a parabola, the areas described by these two bodies in the same time will be as unity to the square root of two, which forms such a connexion between the motion of comets and planets, that, by Kepler's law, the ratio of the areas described during the same time by the comet and the earth may be found; so that the place of a comet at any time in its parabolic orbit, estimated from the instant of its passage at the perihelion, may be computed. But it is a problem of very great difficulty to determine all the other elements of parabolic motion—namely, the comet's perihelion distance, or shortest distance from the sun, estimated in parts of the mean distance of the earth from the sun; the longitude of the perihelion; the inclination of the orbit on the plane of the ecliptic; and the longitude of the ascending node. Three observed longitudes and latitudes of a comet are sufficient for computing the approximate values of these quantities; but an accurate estimation of them can only be obtained by successive corrections from a number of observations, distant from one another. When the motion of a comet is retrograde, the place of the ascending node is exactly opposite to what it is when the motion is direct; hence the place of the ascending node, together with the direction of the comet's motion, show whether the inclination of the orbit is on the north or south side of the plane of the ecliptic. If the motion be direct, the inclination is on the north side; if retrograde, it is on the south side.

The identity of the elements is the only proof of the return of a comet to our system. Should the elements of a new comet be the same, or nearly the same, with those of any one previously known, the probability of the identity of the two bodies is very great, since the similarity extends to no less than four elements, every one of which is capable of an infinity of variations. But even if the orbit be determined with all the accuracy the case admits of, it may be difficult, or even impossible, to recognise a comet on its return, because its orbit would be very much changed if it passed near any of the large planets of this, or of any other system, in consequence of their disturbing energy, which would be very great on bodies of so rare a nature. Halley computed the elements of the orbit of a comet that appeared in the year 1682, which agreed so nearly with those of the comets of 1607 and 1531, that he concluded it to be the same body returning to the sun, at intervals of about seventy-five years. He consequently predicted its re-appearance in the year 1758, or in the beginning of 1759. Science was not sufficiently advanced in the time of Halley, to enable him to determine the perturbations this comet might experience; but Clairaut computed that it would be retarded in its motion a hundred days by the attraction of Saturn, and 518 by that of Jupiter, and consequently, that it would pass its perihelion about the middle of April, 1759, requiring 618 days more to arrive at that point than in its preceding revolution. This, however, he considered only to be an approximation, and that it might be thirty days more or less: the return of the comet on the 12th of March, 1759, proved the truth of the prediction. MM. Damoiseau and Pontecoulant have ascertained that this comet will return either on the 4th or the 7th of November, 1835; the difference of three days in their computations arises from their having employed different values for the masses of the planets. This is the first comet whose periodicity has been established; it is also the first whose elements have been determined from observations made in Europe, for although the comets which appeared in the years 240, 539, 565, and 837, are the most ancient whose orbits have been traced, their elements were computed from Chinese observations.

By far the most curious and interesting instance of the disturbing action of the great bodies of our system is found in the comet of 1770. The elements of its orbit, determined by Messier, did not agree with those of any comet that had hitherto been computed, yet Lexel ascertained that it described an ellipse about the sun, whose major axis was only equal to three times the length of the diameter of the terrestrial orbit, and consequently that it must return to the sun at intervals of five years and a half. This result was confirmed by numerous observations, as the comet was visible through an arc of 170°; yet this comet had never been observed before the year 1770, nor has it ever again been seen, though very brilliant The disturbing action of the larger planets afford a solution of this anomaly, as Lexel ascertained that in

1767 the comet must have passed Jupiter at a distance less than the fifty-eighth part of its distance from the sun, and that in 1779 it would be 500 times nearer Jupiter than the sun; consequently the action of the sun on the comet would not be the fiftieth part of what it would experience from Jupiter, so that Jupiter became the primum mobile. Assuming the orbit to be such as Lexel had determined in 1770, La Place found that the action of Jupiter, previous to the year 1770, had so completely changed the form of it, that the comet which had been invisible to us before 1770, was then brought into view, and that the action of the same planet producing a contrary effect, has, subsequently to that year, removed it, probably for ever, from our system. This comet might have been seen from the earth in 1776, had its light not been eclipsed by that of the sun.

Besides Halley's comet, two others are now proved to form part of our system; that is to say, they return to the sun at intervals, one of 1207 days, and the other of $6\frac{3}{4}$ years, nearly. The first, generally called Encke's comet, or the comet of the short period, was first seen by MM. Messier and Mechain in 1786, again by Miss Herschel in 1795, and its returns in the years 1805 and 1819 were observed by other astronomers, under the impression that all four were different bodies; however, Professor Encke not only proved their identity, but determined the circumstances of the comet's motion. Its reappearance in the years 1825, 1828, and 1832 accorded with the orbit assigned by M. Encke, who thus established the length of its period to be 1207 days, nearly. This comet is very small, of feeble light, and invisible to the naked eye, except under very favourable circumstances, and in particular positions; it has no tail, it revolves in an ellipse of great excentricity inclined at an angle of 13° 22′ to the plane of the ecliptic, and is subject to considerable perturbations from the attraction of the planets. It has already been mentioned, that among the many perturbations to which the planets are liable, their mean motions, and, therefore, the major axes of their orbits, experience no change; while, on the contrary, the mean motion of the moon is accelerated from age to age, a circumstance at first attributed to the resistance of an etherial medium pervading space, but subsequently proved to arise from the secular diminution of the excentricity of the terrestrial orbit. Although the resistance of such a medium has not hitherto been perceived in the motions of such dense bodies as the planets and satellites, its effects on the revolutions of the two small periodic comets hardly leave a doubt of its existence. From the numerous observations that have been made on each return of the comet of the short period, the elements have been computed with great accuracy on the hypothesis of its moving in vacuo; its perturbations occasioned by the disturbing action of the planets have been determined; and after every thing that could influence its motion had been duly considered, M. Encke found that an acceleration of about two days on each revolution has taken place in its mean motion, precisely similar to that which would be occasioned by the resistance of an etherial medium; and as it cannot be attributed to a cause like that which produces the acceleration of the moon, it must be concluded that the celestial bodies do not perform their revolutions in an absolute void, and that although the medium be too rare to have a sensible effect on the masses of the planets and satellites, it nevertheless has a considerable influence on so rare a body as a comet. Contradictory as it may seem, that the motion of a body should be accelerated by the resistance of an etherial medium, the truth becomes evident if it be considered that both planets and comets are retained in their orbits by two forces which exactly balance one another; namely, the centrifugal force producing the velocity in the tangent, and the attraction of the gravitating force directed to the centre of the sun. If one of these forces be diminished by any cause, the other will be proportionally increased. Now, the necessary effect of a resisting medium is to diminish the tangential velocity, so that the balance is destroyed, gravity preponderates, the body descends towards the sun till equilibrium is again restored between the two forces; and as it then describes a smaller orbit, it moves with increased velocity. Thus, the resistance of an etherial medium actually accelerates the motion of a body, but as the resisting force is confined to the plane of the orbit it has no influence whatever on the inclination of the orbit, or on the place of the nodes. The other comet belonging to our system, which returns to its perihelion after a period of 6¾ years, has been accelerated in its motion by a whole day during its last revolution, which puts the existence of ether beyond a doubt, and forms a strong presumption in corroboration of the undulating theory of light. The comet in question was discovered by M. Biela at Johannisberg on the 27th of February, 1826, and ten days afterwards it was seen by M. Gambart at Marseilles, who computed its parabolic elements, and found that they agreed with those of the comets which had appeared in the years 1789 and 1795, whence he concluded them to be the same body moving in an ellipse, and accomplishing its revolution in 2460 days. The perturbations of this comet were computed by M. Damoiseau, who predicted that it would cross the plane of the ecliptic on the 29th of October, 1832, a little before midnight, at a point nearly 18484 miles within the earth's orbit; and as M. Olbers, of Bremen, in 1805, had determined the radius of the comet's head to be about 21136 miles, it was evident that its nebulosity would envelope a portion of the earth's orbit, a circumstance which caused great alarm in France, and not altogether without reason, for if any disturbing cause had delayed the arrival of the comet for one month, the earth must have passed through its head. M. Arago dispelled their fears by the excellent treatise on comets which appeared in the Annuaire of 1832, where he proves that, as the earth would never be nearer the comet than 24800000 British leagues, there could be no danger of collision.

If a comet were to impinge on the earth, so as to destroy its centrifugal force, it would fall to the sun in 64½ days. What the earth's primitive velocity may have been it is impossible to say; therefore a comet may have given it a shock without changing the axis of rotation, but only destroying part of its tangential velocity, so as to diminish the size of the orbit, a thing by no means impossible, though highly improbable; at all events, there is no proof that such has been the case; and it is manifest that the axis of the earth's rotation has not been changed, because, as there is no resistance, the libration would to this day be evident in the variation it must have occasioned in the terrestrial latitudes. Supposing the nucleus of a comet to have a diameter only equal to the fourth part of that of the earth, and that its perihelion is nearer to the sun than we are ourselves, its orbit being otherwise unknown, M. Arago has computed that the probability of the earth receiving a shock from it is only one in 281 millions, and that the chance of our coming in contact with its nebulosity is about ten or twelve times greater. But in general comets are so rare, that it is likely they would not do much harm if they were to impinge; and even then the mischief would probably be local, and the equilibrium soon restored, provided there was only a gaseous or very small nucleus. It is, however, more probable that the earth would only be deflected a little from its course by the approach of a comet, without being touched by it. The comets that seem to have come nearest to the earth were that of the year 837, which remained four days within less than 1240000 leagues from our orbit; that of 1770, which approached within about six times the distance of the moon. The celebrated comet of 1680 also came very near to us; and the comet whose period is $6\frac{3}{4}$ years was ten times nearer the earth in 1805 than in 1832, when it caused so much alarm.

Comets, when in or near their perihelion, move with prodigious velocity. That of 1680 appears to have gone half round the sun in ten hours and a half, moving at the rate of 880000 miles an hour. If its enormous centrifugal force had ceased when passing its perihelion, it would have fallen to the sun in about three minutes, as it was then only 147000 miles from his surface. So near the sun, it would be exposed to a heat 27500 times greater than that received by the earth; and as the sun's heat is supposed to be in proportion to the intensity of his light, it is probable that a degree of heat so very intense would be sufficient to convert into vapour every terrestrial substance with which we are acquainted. At the perihelion distance the sun's diameter would be seen from the comet under an angle of 73°, so that the sun, viewed from the comet, would nearly cover the whole extent of the heavens from the horizon to the zenith; and as this comet is presumed to have a period of 575 years, the major axis of its orbit must be so great, that at the aphelion the sun's diameter would only subtend an angle of about fourteen seconds, which is not so great as half the diameter of Mars appears to us when in opposition. The sun would consequently impart no heat, so that the comet would then be exposed to the temperature of the etherial regions, which is 58° below the zero point of Fahrenheit. A body so rare as the comet, and moving with such velocity, must have met with great resistance from the dense atmosphere of the sun, while passing so near his surface at its perihelion. The centrifugal force must consequently have been diminished, and the sun's attraction proportionally augmented, so that it must have come nearer to the sun in 1680 than in its preceding revolution, and would subsequently describe a smaller orbit. As this diminution of its orbit will be repeated at each revolution, the comet will infallibly end by falling on the surface of the sun, unless its course be changed by the disturbing influence of some large body in the unknown expanse of creation. Our ignorance of the actual density of the sun's atmosphere, of the density of the comet, and of the period of its revolution, renders it impossible to form any idea of the number of centuries which must elapse before this singular event takes place.

But this is not the only comet threatened with such a catastrophe; Encke's, and that discovered by M. Biela, are both slowly tending to the same fate. By the resistance of the ether, they will perform each revolution nearer and nearer to the sun, till at last they will be precipitated on his surface. The same cause may affect the motions of the planets, and be ultimately the means of destroying the solar system; but, as Sir John Herschel observes, they could hardly all revolve in the same direction round the sun for so many ages without impressing a corresponding motion on the etherial fluid, which may preserve them from the accumulated effects of its resistance. Should this material fluid revolve about the sun like a vortex, it will accelerate the revolutions of such comets as have direct motions, but it will retard those that have retrograde motions.

Though already so well acquainted with the motions of comets, we know nothing of their physical constitution. A vast number, especially of telescopic comets, are only like clouds or masses of vapour often without tails. Such were the comets which appeared in the years 1795, 1797, and 1798; but the head commonly consists of a mass of light, like a planet surrounded by a very transparent atmosphere, the whole, viewed with a telescope, being so diaphanous, that the smallest star may be seen even through the densest part of the nucleus; and in general their masses, when they have any, are so minute that they have no sensible diameter, like that of the comet of 1811, which appeared to Sir Wm. Herschel like a luminous point in the middle of the nebulous matter. The nuclei, which seem to be formed of the denser strata of that nebulous matter in successive coatings, are often of great magnitude; those of the comets which came to the sun in the years 1799 and 1807 had nuclei whose diameters measured 180 and 275 leagues respectively, and the second comet of 1811 had a nucleus 1350 leagues in diameter.

The nebulosity immediately round the nucleus is so diaphanous that it gives little light; but at a small distance the nebulous matter becomes suddenly brilliant, so as to look like a bright ring round the body. Sometimes there are as many as two or three of these luminous concentric rings separated by dark intervals, but they are generally incomplete on the side next the tail. In the comet of 1811, the luminous ring was 12400 leagues thick, and the distance between its interior surface and the centre of the nucleus was as much as 14880 leagues. The thickness of these bright diaphanous coatings in the comets of 1807 and 1799 were 14880 and 9920 leagues respectively. The transit of a comet over the sun would afford the best information with regard to the nature of the nuclei. It was computed that such an event was to take place in the year 1827; unfortunately the sun was hid by clouds from the British astronomers, but it was examined at Viviers and at Marseilles, at the time the comet must have been projected on its disc, but no spot or cloud was to be seen.

The tails of comets proceed from the head in two streams of light somewhat like that of the aurora; these in most cases unite at a greater or less distance from the nucleus, and are generally situate in the planes of their orbits; they follow the comets in their descent towards the sun, but precede them in their return with a small degree of curvature, probably owing to the resistance of the ether, but their extent and form must vary in appearance according to the positions of their orbits with regard to the ecliptic. In some cases, the tail has been at right angles to the line joining the sun and comet. They are generally of enormous lengths,—the comet of 1811 had a tail no less than 34 millions of leagues in length, and those which appeared in the years 1618, 1680, and 1769, had tails which extended respectively over 104, 90, and 97 degrees of space; consequently, when the heads of these comets were set, a portion of the extremity of their tails was still in the zenith. Sometimes the tail is divided into several branches, like the comet of 1744, which had six, separated by dark intervals, each of them about 4° broad, and from 30° to 44° long. The tails do not attain their full magnitude till the comet has left the sun. When these bodies first appear, they resemble round films of vapour with little or no tail; as they approach the sun, they increase in brilliancy, and their tail in length, till they are lost in his rays; and it is not till they emerge from the sun's more vivid light that they assume their full splendour. They then gradually decrease by the same degrees; their tails diminish and disappear nearly or altogether before the comet is beyond the sphere of telescopic vision. Many comets have no tails at all, as, for example, Encke's comet and that discovered by M. Biela, both of which are small and insignificant objects. The comets which appeared in the years 1585, 1763, and 1682, were also without tails, though the latter is recorded to have been as bright as Jupiter. The matter of the tail must be extremely buoyant to precede a body moving with such velocity; indeed the rapidity of its ascent can only be accounted for by the fervent heat of the sun. Immediately after the great comet of 1680 had passed its perihelion, its tail was 20000000 leagues in length, and was projected from the comet's head in the short space of two days. A body of such extreme tenuity as a comet is most likely incapable of an attraction powerful enough to recall matter sent to such an enormous distance; it is therefore, in all probability, scattered in space, which may account for the rapid decrease observed in the tails of comets every time they return to their perihelia.

It is remarkable that, although the tails of comets increase in length as they approach their perihelia, there is reason to believe that the real diameter of the nebulous matter or nucleus contracts on coming near the sun, and expands rapidly on leaving him. Hevelius first observed this phenomenon, which Encke's comet has exhibited in a very extraordinary degree. On the 28th of October, 1828, this comet was about three times as far from the sun as it was on the 24th of December, yet at the first date its apparent diameter was twenty-five times greater than at the second, the decrease being progressive. M. Valz attributes the circumstance to a real condensation of volume from the pressure of the ethereal medium, which increases most rapidly in density towards the surface of the sun, and forms an extensive atmosphere around him. Sir John Herschel, on the contrary, conjectures that it may be owing to the alternate conversion of evaporable materials in the upper regions of a transparent atmosphere into the states of visible cloud and invisible gas by the effects of heat and cold. Not only the tails, but the nebulous part of comets diminishes every time they return to their perihelia; after frequent returns they ought to lose it altogether, and present the appearance of a fixed nucleus: this ought to happen sooner to comets of short periods. La Place supposes that the comet of 1682 must be approaching rapidly to that state. Should the substances be altogether, or even to a great degree, evaporated, the comet would disappear for ever. Possibly comets may have vanished from our view sooner than they would otherwise have done from this cause.

In those positions of comets where only half of their enlightened hemisphere ought to be seen if they shine by reflected light, they ought to exhibit phases, but even with high magnifying powers none have been detected, though some slight indications are said to have been once observed by Hevelius and La Hire in 1682. In general the light of comets is dull,—that of the comet of 1811 was only equal to the tenth part of the light of the full moon, but some have been brilliant enough to be visible in full daylight, especially the comet of 1744, which was seen without a telescope at one o'clock in the afternoon, while the sun was shining; whence it may be inferred that, although some comets may be altogether diaphanous, others seem to possess a solid mass resembling a planet; but whether they shine by their own or by reflected light has never been satisfactorily made out till now. As light is polarized by reflection at certain angles, it would afford a decisive test, were it not that a body is capable of reflecting light, though it shines by its own; so that it would not be conclusive, even if the light of a comet were polarized light. M. Arago, however, has with great ingenuity discovered a method of ascertaining this point, independent both of phases and polarization.

Since the rays of light diverge from a luminous point, they will be scattered over a greater space as the distance increases, so that the intensity of the light on a screen two feet from the object is four times less than at the distance of one foot; three feet from the object it is nine times less, and so on, decreasing in intensity as the square of the distance increases. As a self-luminous surface consists of an infinite number of luminous points, it is clear that, the greater the extent of surface, the more intense will be the light; whence it may be concluded that the illuminating power of such a surface is proportional to its extent, and decreases inversely as the square of the distance. Notwithstanding this, a self-luminous surface, plane or curved, viewed through a hole in a plate of metal, is of the same brilliancy at all possible distances as long as it subtends a sensible angle, because, as the distance increases, a greater portion comes into view, and as the augmentation of surface is as the square of the diameter of the part seen through the hole, it increases as the square of the distance. Hence, though the number of rays from any one point of the surface which pass through the hole decrease inversely as the square of the distance, yet, as the extent of surface which comes into view increases also in that ratio, the brightness of the object is the same to the eye as long as it has a sensible diameter. For example—Uranus is about nineteen times farther from the sun than we are, so that the sun, seen from that planet, must appear like a star with a diameter of a hundred seconds, and must have the same brilliancy to the inhabitants that he would have to us if viewed through a small circular hole having a diameter of a hundred seconds. For it is obvious that light comes from every point of the sun's surface to Uranus, whereas a very small portion of his disc is visible through the hole; so that extent of surface exactly compensates distance. Since, then, the visibility of a self-luminous object does not depend upon the angle it subtends as long as it is of sensible magnitude, if a comet shines by its own light, it should retain its brilliancy as long as its diameter is of a sensible magnitude; and even after it has lost an apparent diameter, it ought, like the fixed stars, to be visible, and should only vanish in consequence of extreme remoteness. That, however, is far from being the case—comets gradually become dim as their distance increases, and vanish merely from loss of light, while they still retain a sensible diameter, which is proved by observations made the evening before they disappear. It may therefore be concluded that comets shine by reflecting the sun's light. The most brilliant comets have hitherto ceased to be visible when about five times as far from the sun as we are. Most of the comets that have been visible from the earth have their perihelia within the orbit of Mars, because they are invisible when as distant as the orbit of Saturn: on that account there is not one on record whose perihelion is situate beyond the orbit of Jupiter. Indeed, the comet of 1756, after its last appearance, remained five whole years within the ellipse described by Saturn without being once seen. A hundred and forty comets have appeared within the earth's orbit during the last century that have not again been seen. If a thousand years be allowed as the average period of each, it may be computed, by the theory of probabilities, that the whole number which range within the earth's orbit must be 1400; but Uranus being about nineteen times more distant, there may be no less than 11200000 comets that come within the known extent of our system. M. Arago makes a different estimate: he considers that, as thirty comets are known to have their perihelion distance within the orbit of Mercury, if it be assumed that comets are uniformly distributed in space, the number having their perihelion within the orbit of Uranus must be to thirty as the cube of the radius of the orbit of Uranus to the cube of the radius of the orbit of Mercury, which makes the number of comets amount to 3529470; but that number may be doubled if it be considered that, in consequence of day-light, fogs, and great southern declination, one comet out of two is concealed from us. So, according to M. Arago, more than seven millions of comets frequent the planetary orbits.

SECTION XXXVI.

Great as the number of comets appears to be, it is absolutely nothing when compared to the number of the fixed stars. About two thousand only are visible to the naked eye; but when we view the heavens with a telescope, their number seems to be limited only by the imperfection of the instrument. In one hour Sir William Herschel estimated that 50000 stars passed through the field of his telescope, in a zone of the heavens 2° in breadth. This, however, was stated as an instance of extraordinary crowding; but, at an average, the whole expanse of the heavens must exhibit about a hundred millions of fixed stars that come within the reach of telescopic vision.

The stars are classed according to their apparent brightness, and the places of the most remarkable of those visible to the naked eye are ascertained with great precision, and formed into a catalogue, not only for the determination of geographical position by their occultations, but to serve as points of reference for finding the places of comets and other celestial phenomena. The whole number of stars registered amounts to about 15000 or 20000. The distance of the fixed stars is too great to admit of their exhibiting a sensible disc; but, in all probability, they are spherical, and must certainly be so if gravitation pervades all space, which it may be presumed to do, since Sir John Herschel has shown that it extends to the binary systems of stars. With a fine telescope the stars appear like a point of light, their occultations by the moon are therefore instantaneous; their twinkling arises from sudden changes in the refractive power of the air, which would not be sensible if they had discs like the planets. Thus we can learn nothing of the relative distances of the stars from us and from one another by their apparent diameters; but their annual parallax being insensible, shows that we must be one hundred millions of millions of miles at least from the nearest; many of them, however, must be vastly more remote, for of two stars that appear close together, one may be far beyond the other in the depth of space. The light of Sirius, according to the observations of Sir John Herschel, is 324 times greater than that of a star of the sixth magnitude; if we suppose the two to be really of the same size, their distances from us must be in the ratio of 57·3 to 1, because light diminishes as the square of the distance of the luminous body increases.

Nothing is known of the absolute magnitude of the fixed stars, but the quantity of light emitted by many of them shows that they must be much larger than the sun. Dr. Wollaston determined the approximate ratio that the light of a wax candle bears to that of the sun, moon, and stars, by comparing their respective images, reflected from small glass globes filled with mercury, whence a comparison was established between the quantities of light emitted by the celestial bodies themselves. By this method he found that the light of the sun is about twenty millions of millions of times greater than that of Sirius, the brightest, and supposed to be the nearest of the fixed stars. If Sirius had a parallax of half a second, its distance from the earth would be 525481 times the distance of the sun from the earth; and therefore Sirius, placed where the sun is, would appear to us to be 3·7 times as large as the sun, and would give 13·8 times more light: but many of the fixed stars must be infinitely larger than Sirius.

Many stars have vanished from the heavens; the star 42 Virginis seems to be of this number, having been missed by Sir John Herschel on the 9th of May, 1828, and not again found, though he frequently had occasion to observe that part of the heavens. Sometimes stars have all at once appeared, shone with a bright light, and vanished. Several instances of these temporary stars are on record; a remarkable instance occurred in the year 125, which is said to have induced Hipparchus to form the first catalogue of stars. Another star appeared suddenly near α Aquilæ in the year 389, which vanished after remaining for three weeks as bright as Venus. On the 10th of October, 1604, a brilliant star burst forth in the constellation of Serpentarius, which continued visible for a year; and a more recent case occurred in the year 1670, when a new star was discovered in the head of the Swan, which, after becoming invisible, reappeared, and after many variations in light vanished after two years, and has never since been seen. In 1572, a star was discovered in Cassiopeia, which rapidly increased in brightness till it even surpassed that of Jupiter; it then gradually diminished in splendour, and after exhibiting all the variety of tints that indicates the changes of combustion, vanished sixteen months after its discovery without altering its position. It is impossible to imagine anything more tremendous than a conflagration that could be visible at such a distance. It is however suspected that this star may be periodical and identical with the stars which appeared in the years 945 and 1264. There are probably many stars which alternately vanish and reappear among the innumerable multitudes that spangle the heavens, the periods of thirteen have already been pretty well ascertained. Of these the most remarkable is the star Omicron in the constellation Cetus. It appears about twelve times in eleven years, and is of variable brightness, sometimes appearing like a star of the second magnitude; but it neither always arrives at the same lustre, nor does it increase or diminish by the same degrees. According to Hevelius, it did not appear at all for four years. γ Hydræ also vanishes and reappears every 494 days, and a very singular instance of periodicity is given by Sir John Herschel in the star Algol or β Persei, which is described as retaining the size of a star of the second magnitude for two days and fourteen seconds; it then suddenly begins to diminish in splendour, and in about three hours and a half is reduced to the size of a star of the fourth magnitude: it then begins again to increase, and in three hours and a half more regains its usual brightness, going through all these vicissitudes in two days, twenty hours, and forty-eight minutes. The cause of the variations in most of the periodical stars is unknown, but, from the changes of Algol, M. Goodricke has conjectured that they may be occasioned by the revolution of some opaque body, coming between us and the star, obstructing part of its light. Sir John Herschel is struck with the high degree of activity evinced by these changes in regions where, 'but for such evidences, we might conclude all to be lifeless.' He observes that our own sun requires nine times the period of Algol to perform a revolution on its own axis; while, on the other hand, the periodic time of an opaque revolving body sufficiently large to produce a similar temporary obscuration of the sun, seen from a fixed star, would be less than fourteen hours.

Many thousands of stars that seem to be only brilliant points, when carefully examined are found to be in reality systems of two or more suns, some revolving about a common centre. These binary and multiple stars are extremely remote, requiring the most powerful telescopes to show them separately. The first catalogue of double stars, in which their places and relative positions are determined, was accomplished by the talents and industry of Sir William Herschel, to whom astronomy is indebted for so many brilliant discoveries, and with whom the idea of their combination in binary and multiple systems originated — an idea completely established by his own observations, recently confirmed by those of his son. The motions of revolution of many round a common centre have been ascertained, and their periods determined with considerable accuracy. Some have, since their first discovery, already accomplished nearly a whole revolution, and one, η Coronæ, is actually considerably advanced in its second period. These interesting systems thus present a species of sidereal chronometer, by which the chronology of the heavens will be marked out to future ages by epochs of their own, liable to no fluctuations from planetary disturbances, such as obtain in our system."

In observing the relative position of the stars of a binary system, the distance between them, and also the angle of position, that is, the angle which the meridian or a parallel to the equator makes with the line joining the two stars are measured. The accuracy of each result depends upon taking the mean of a great number of the best observations, and eliminating error by mutual comparison. The distances between the stars are so minute that they cannot be measured with the same accuracy as the angles of position; therefore, to determine the orbit of a star independently of the distance, it is necessary to assume, as the most probable hypothesis, that the stars are subject to the law of gravitation, and consequently, that one of the two stars revolves in an ellipse about the other, supposed to be at rest, though not necessarily in the focus. A curve is thus constructed graphically by means of the angles of position and the corresponding times of observation. The angular velocities of the stars are obtained by drawing tangents to this curve at stated intervals, whence the apparent distances, or radii vectores, of the revolving star become known for each angle of position; because, by the laws of elliptical motion, they are equal to the square roots of the apparent angular velocities. Now that the angles of position estimated from a given line, and the corresponding distances of the two stars, are known, another curve may be drawn, which will represent on paper the actual orbit of the star projected on the visible surface of the heavens; so that the elliptical elements of the true orbit and its position in space may be determined by a combined system of measurements and computation. But as this orbit has been obtained on the hypothesis that gravitation prevails in these distant regions, which could not be known *à priori*, it must be compared with as many observations as can be obtained, to ascertain how far the computed ellipse agrees with the curve actually described by the star.

By this process Sir John Herschel has discovered that several of these systems of stars are subject to the same laws of motion with our system of planets: he has determined the elements of their elliptical orbits, and computed the periods of their revolution. One of the stars of γ Virginis revolves about the other in 629 years; the periodic time of σ Coronæ is 287 years; that of Castor is 253 years; that of ε Bootes is 1600; that of 70 Ophiuci is ascertained by M. Savary to be 80 years; and Professor Encke has shown that the revolution of ξ Ursæ is completed in 58 years. The two first of these stars are approaching their perihelia,—γ Virginis will arrive at it on the 18th of August, 1834, and Castor some time in 1855. The actual proximity of the two component stars in each case will then be extreme, and the apparent angular velocity so great, that, in the case of γ Virginis, an angle of 68° may be described in a single year. σ Coronæ will also attain its perihelion about 1835. Sir John Herschel, Sir James South, and Professor Struve of Dorpat, have increased Sir William Herschel's original catalogue to more than 3000, of which thirty or forty are known to form revolving or binary systems, and Mr. Dunlop has formed a catalogue of 253 double stars in the southern hemisphere. The motion of Mercury is more rapid than that of any other planet, being at the rate of 107000 miles in an hour; the perihelion velocity of the comet of 1680 was no less than 880000 miles an hour; but if the two stars of ξ Ursæ be as remote from one another as the nearest fixed star is from the sun, the velocity of the revolving stars must exceed imagination. The discovery of the elliptical motion of the double stars excites the highest interest, since it shows that gravitation is not peculiar to our system of planets, but that systems of suns in the far distant regions of the universe are also obedient to its laws.

Possibly, among the multitudes of small stars, whether double or insulated, some may be found near enough to exhibit distinct parallactic motions, arising from the revolution of the earth in its orbit. Of two stars apparently in close approximation, one may be far behind the other in space. These may seem near to one another when viewed from the earth in one part of its orbit, but may separate widely when seen from the earth in another position, just as two terrestrial objects appear to be one when viewed in the same straight line, but separate as the observer changes his position. In this case the stars would not have real, but only apparent motion. One of them would seem to oscillate annually to and fro in a straight line on each side of the other—a motion which could not be mistaken for that of a binary system, where one star describes an ellipse about the other. Such parallax does not yet appear to have been made out, so that the actual distance of the stars is still a matter of conjecture.

The double stars are of various hues, but most frequently exhibit the contrasted colours. The large star is generally yellow, orange, or red; and the small star blue, purple, or green. Sometimes a white star is combined with a blue or purple, and more rarely a red and white are united. In many cases, these appearances are due to the influences of contrast on our judgment of colours. For example, in observing a double star, where the large one is a full ruby-red or almost blood-colour, and the small one a fine green, the latter loses its colour when the former is hid by the cross wires of the telescope. But there are a vast number of instances where the colours are too strongly marked to be merely imaginary. Sir John Herschel observes in one of his papers in the Philosophical Transactions, as a very remarkable fact, that, although red stars are common enough, no example of an insulated blue, green, or purple one has yet been produced.

Besides the revolutions about one another, some of the binary systems are carried forward in space by a motion common to both stars, towards some unknown point in the firmament. The two stars of 61 Cygni, which are nearly equal, and have remained at the distance of about 15″ from each other for fifty years, have changed their place in the heavens during that period, by a motion which for ages must appear uniform and rectilinear: because, even if the path be curved, so small a portion of it must be sensibly a straight line to us. Multitudes of the single stars also have proper motions, yet so minute that that of μ Cassiopeiæ, which is only 3″·74 annually, is the greatest yet observed; but the enormous distances of the stars make motions appear small to us which are in reality very great. Sir William Herschel conceived that, among many irregularities, the motions of the stars have a general tendency towards a point diametrically opposite to that occupied by the star ζ Herculis, which he attributed to a motion of the solar system in the contrary direction. Should this really be the case, the stars, from the effects of perspective alone, would seem to diverge in the direction to which we are tending, and would apparently converge in the space we leave, and there would be a regularity in these apparent motions which would in time be detected; but if the solar system and the whole of the stars visible to us be carried forward in space by a motion common to all, like ships drifting in a current, it would be impossible for us, who move with the rest, to ascertain its direction. There can be no doubt of the progressive motion of the sun and many of the stars, but sidereal astronomy is not far enough advanced to determine what relations these bear to one another.

The stars are scattered very irregularly over the firmament. In some places they are crowded together, in others thinly dispersed. A few groups more closely condensed form very beautiful objects even to the naked eye, of which the Pleiades and the constellation Coma Berenices are the most striking examples; but the greater number of these clusters of stars appear to unassisted vision like thin white clouds or vapours: such is the milky way, which, as Sir William Herschel has proved, derives its brightness from the diffused light of the myriads of stars that form it. Most of them are extremely small on account of their enormous distances, and they are so numerous that, according to his estimation, no fewer than 50000 passed through the field of his telescope in the course of one hour in a zone 2° broad. This singular portion of the heavens, constituting part of our firmament, consists of an extensive stratum of stars, whose thickness is small compared with its length and breadth; the earth is placed about midway between its two surfaces, near the point where it diverges into two branches. Many clusters of stars appear like white clouds or round comets without tails, either to unassisted vision or with ordinary telescopes; but with powerful instruments Sir John Herschel describes them as conveying the idea of a globular space filled full of stars insulated in the heavens, and constituting a family or society apart from the rest, subject only to its own internal laws. To attempt to count the stars in one of these globular clusters, he says, would be a vain task,—that they are not to be reckoned by hundreds,—and, on a rough computation, it appears that many clusters of this description must contain ten or twenty thousand stars compacted and wedged together in a round space whose area is not more than a tenth part of that covered by the moon; so that its centre, where the stars are seen projected on each other, is one blaze of light. If each of these stars be a sun, and if they be separated by intervals equal to that which separates our sun from the nearest fixed star, the distance which renders the whole cluster barely visible to the naked eye must be so great, that the existence of this splendid assemblage can only be known to us by light which must have left it at least a thousand years ago. Occasionally these clusters are so irregular and so undefined in their outline as merely to suggest the idea of a richer part of the heavens. They contain fewer stars than the globular clusters, and sometimes a red star forms a conspicuous object among them. These Sir William Herschel regarded as the rudiments of globular clusters in a less advanced state of condensation, but tending to that form by their mutual attraction.

Multitudes of nebulous spots are to be seen on the clear vault of heaven which have every appearance of being clusters like those described, but are too distant to be resolved into stars by the most excellent telescopes. This nebulous matter exists in vast abundance in space. No fewer than 2000 nebulæ and clusters of stars were observed by Sir William Herschel, whose places have been computed from his observations, reduced to a common epoch, and arranged into a catalogue in order of right ascension by his sister, Miss Caroline Herschel, a lady so justly eminent for astronomical knowledge and discovery. Six or seven hundred nebulæ have already been ascertained in the southern hemisphere; of these the magellanic clouds are the most remarkable. The nature and use of this matter, scattered over the heavens in such a variety of forms, is involved in the greatest obscurity. That it is a self-luminous, phosphorescent, material substance, in a highly dilated or gaseous state, but gradually subsiding by the mutual gravitation of its particles into stars and sidereal systems, is the hypothesis which seems to be most generally received; but the only way that any real knowledge on this mysterious subject can be obtained is by the determination of the form, place, and present state of each individual nebula; and a comparison of these with future observations will show generations to come the changes that may now be going on in these supposed rudiments of future systems. With this view, Sir John Herschel began in the year 1825 the arduous and pious task of revising his illustrious father's observations, which he finished a short time before he sailed for the Cape of Good Hope, in order to disclose the mysteries of the southern hemisphere, because he considers our firmament to be exhausted till farther improvements in the telescope shall enable astronomers to penetrate deeper into space. In a truly splendid paper read before the Royal Society on the 21st of November, 1833, he gives the places of 2500 nebulæ and clusters of stars. Of these, 500 are new,—the rest he mentions with peculiar pleasure as having been most accurately determined by his father. This work is the more extraordinary, as, from bad weather, fogs, twilight, and moonlight, these shadowy appearances are not visible, at an average, above thirty nights in the year.

The nebulæ have a great variety of forms. Vast multitudes are so faint as to be with difficulty discerned at all till they have been for some time in the field of the telescope, or are just about to quit it. Many present a large ill-defined surface, in which it is difficult to say where the centre of the greatest brightness is. Some cling to stars like wisps of cloud; others exhibit the wonderful appearance of an enormous flat ring seen very obliquely, with a lenticular vacancy in the centre. A very remarkable instance of an annular nebula is to be seen exactly half-way between β and γ Lyræ. It is elliptical in the ratio of 4 to 5, is sharply defined, the internal opening occupying about half the diameter. This opening is not entirely dark, but filled up with a faint hazy light, aptly compared by Sir John Herschel to fine gauze stretched over a hoop. Two are described as most amazing objects:—one like a dumb-bell or hour-glass of bright matter, surrounded by a thin hazy atmosphere, so as to give the whole an oval form, or the appearance of an oblate spheroid. This phenomenon bears no resemblance to any known object. The other consists of a bright round nucleus, surrounded at a distance by a nebulous ring split through half its circumference, and having the split portions separated at an angle of 45° each to the plane of the other. This nebula bears a strong similitude to the milky way, and suggested to Sir John Herschel the idea of 'a brother system bearing a real physical resemblance and strong analogy of structure to our own.' It appears that double nebulæ are not unfrequent, exhibiting all the varieties of distance, position, and relative brightness with their counterparts the double stars. The rarity of single nebulæ as large, faint, and as little condensed in the centre as these, makes it extremely improbable that two such bodies should be accidentally so near as to touch, and often in part to overlap each other as these do. It is much more likely that they constitute systems; and if so, it will form an interesting subject of future inquiry to discover whether they possess orbitual motion round one another.

Stellar nebulæ form another class. These have a round or oval shape, increasing in density towards the centre. Sometimes the matter is so rapidly condensed as to give the whole the appearance of a star with a blur, or like a candle shining through horn. In some instances the central matter is so highly and suddenly condensed, so vivid and sharply defined, that the nebula might be taken for a bright star surrounded by a thin atmosphere. Such are nebulous stars. The zodiacal light, or lenticular-shaped atmosphere of the sun, which may be seen extending beyond the orbits of Mercury and Venus soon after sunset in the months of April and May, is supposed to be a condensation of the ethereal medium by his attractive force, and seems to place our sun among the class of stellar nebulæ. The stellar nebulæ and nebulous stars assume all degrees of ellipticity. Not unfrequently they are long and narrow, like a spindle-shaped ray, with a bright nucleus in the centre. The last class are the planetary nebulæ. These bodies have exactly the appearance of planets, with sensibly round or oval discs, sometimes sharply terminated, at other times hazy and ill defined. Their surface, which is blue or bluish-white, is equable or slightly mottled, and their light occasionally rivals that of the planets in vividness. They are generally attended by minute stars which give the idea of accompanying satellites. These nebulæ are of enormous dimensions. One of them, near ν Aquarii, has a sensible diameter of about 20″, and another presents a diameter of 12″. Sir John Herschel has computed that, if these objects be as far from us as the stars, their real magnitude must, even on the lowest estimation, be such as would fill the orbit of Uranus. He concludes that, if they be solid bodies of a solar nature, their intrinsic splendour must be greatly inferior to that of the sun, because a circular portion of the sun's disc, subtending an angle of 20″, would give a light equal to that of a hundred full moons, while, on the contrary, the objects in question are hardly, if at all, visible to the naked eye. From the uniformity of the discs of the planetary nebulæ, and their want of apparent condensation, he presumes that they may be hollow shells, only emitting light from their surfaces.

The existence of every degree of ellipticity in the nebulæ—from long lenticular rays to the exact circular form,—and of every shade of central condensation—from the slightest increase of density to apparently a solid nucleus,—may be accounted for by supposing the general constitution of these nebulæ to be that of oblate spheroidal masses of every degree of flatness, from the sphere to the disc, and of every variety in their density and ellipticity towards the centre. It would be erroneous however to imagine, that the forms of these systems are maintained by forces identical with those already described, which determine the form of a fluid mass in rotation; because, if the nebulæ be only clusters of separate stars, as in the greater number of cases there is every reason to believe them to be, no pressure can be propagated through them. Consequently, since no general rotation of such a system as one mass can be supposed, it may be conceived to be a quiescent form, comprising within its limits an indefinite multitude of stars, each of which may be moving in an orbit about the common centre of the whole, in virtue of a law of internal gravitation resulting from the compound gravitation of all its parts. Sir John Herschel has proved that the existence of such a system is not inconsistent with the law of gravitation under certain conditions.

The distribution of the nebulæ over the heavens is even more irregular than that of the stars. In some places they are so crowded together as scarcely to allow one to pass through the field of the telescope before another appears, while in other parts hours elapse without a single nebula occurring in the zone under observation. They are in general only to be seen with the very best telescopes, and are most abundant in a zone whose general direction is not far from the hour circles 0^h and 12^h, and which crosses the milky way nearly at right angles. Where that zone crosses the constellations Virgo, Coma Berenices, and the Great Bear, they are to be found in multitudes.

Such is a brief account of the discoveries contained in Sir John Herschel's paper, which, for sublimity of views and patient investigation, has not been surpassed in any age or country. To him and to Sir William Herschel is due almost all that is known of sidereal astronomy; and in the inimitable works of that highly-gifted father and son, the reader will find this subject treated of in a style altogether worthy of it, and of them.

So numerous are the objects which meet our view in the heavens, that we cannot imagine a part of space where some light would not strike the eye;—innumerable stars, thousands of double and multiple systems, clusters in one blaze with their tens of thousands of stars, and the nebulæ amazing us by the strangeness of their forms and the incomprehensibility of their nature, till at last, from the imperfection of our senses, even these thin and airy phantoms vanish in the distance. If such remote bodies shine by reflected light, we should be unconscious of their existence; each star must then be a sun, and may be presumed to have its system of planets, satellites, and comets, like our own; and, for aught we know, myriads of bodies may be wandering in space unseen by us, of whose nature we can form no idea, and still less of the part they perform in the economy of the universe; nor is this an unwarranted presumption; many such do come within the sphere of the earth's attraction, are ignited by the velocity with which they pass through the atmosphere, and are precipitated with great violence on the earth. The fall of meteoric stones is much more frequent than is generally believed; hardly a year passes without some instances occurring, and if it be considered that only a small part of the earth is inhabited, it may be presumed that numbers fall in the ocean or on the uninhabited part of the land, unseen by man. They are sometimes of great magnitude; the volume of several has exceeded that of the planet Ceres, which is about 70 miles in diameter. One which passed within 25 miles of us was estimated to weigh about 600000 tons, and to move with a velocity of about 20 miles in a second,—a fragment of it alone reached the earth. The obliquity of the descent of meteorites, the peculiar substances they are composed of, and the explosion accompanying their fall, show that they are foreign to our system. Luminous spots, altogether independent of the phases, have occasionally appeared on the dark part of the moon; these have been ascribed to the light arising from the eruption of volcanos; whence it has been supposed that meteorites have been projected from the moon by the impetus of volcanic eruption. It has even been computed that, if a stone were projected from the moon in a vertical line, with an initial velocity of 10992 feet in a second,—more than four times the velocity of a ball when first discharged from a cannon,—instead of falling back to the moon by the attraction of gravity, it would come within the sphere of the earth's attraction, and revolve about it like a satellite. These bodies, impelled either by the direction of the primitive impulse, or by the disturbing action of the sun, might ultimately penetrate the earth's atmosphere, and arrive at its surface. But from whatever source meteoric stones may come, it seems highly probable that they have a common origin, from the uniformity—we may almost say identity—of their chemical composition.

SECTION XXXVII.

The known quantity of matter bears a very small proportion to the immensity of space. Large as the bodies are, the distances which separate them are immeasurably greater; but as design is manifest in every part of creation, it is probable that, if the various systems in the universe had been nearer to one another, their mutual disturbances would have been inconsistent with the harmony and stability of the whole. It is clear that space is not pervaded by atmospheric air, since its resistance would, long ere this, have destroyed the velocity of the planets; neither can we affirm it to be a void, since it is replete with ether, and traversed in all directions by light, heat, gravitation, and possibly by influences whereof we can form no idea.

Whatever the laws may be that obtain in the more distant regions of creation, we are assured that one alone regulates the motions not only of our own system, but also the binary systems of the fixed stars; and as general laws form the ultimate object of philosophical research, we cannot conclude these remarks without considering the nature of gravitation —that extraordinary power whose effects we have been endeavouring to trace through some of their mazes. It was at one time imagined that the acceleration in the moon's mean motion was occasioned by the successive transmission of the gravitating force; but it has been proved that, in order to produce this effect, its velocity must be about fifty millions of times greater than that of light, which flies at the rate of 200000 miles in a second: its action, even at the distance of the sun, may therefore be regarded as instantaneous; yet so remote are the nearest of the fixed stars, that it may be doubted whether the sun has any sensible influence on them.

The curves in which the celestial bodies move by the force of gravitation are only lines of the second order; the attraction of spheroids, according to any other law of force than that of gravitation, would be much more complicated; and as it is easy to prove that matter might have been moved according to an infinite variety of laws, it may be concluded that gravitation must have been selected by Divine Wisdom out of an infinity of others, as being the most simple, and that which gives the greatest stability to the celestial motions.

It is a singular result of the simplicity of the laws of nature, which admit only of the observation and comparison of ratios, that the gravitation and theory of the motions of the celestial bodies are independent of their absolute magnitudes and distances; consequently, if all the bodies of the solar system, their mutual distances, and their velocities, were to diminish proportionally, they would describe curves in all respects similar to those in which they now move; and the system might be successively reduced to the smallest sensible dimensions, and still exhibit the same appearances. We learn by experience that a very different law of attraction prevails when the particles of matter are placed within inappreciable distances from each other, as in chemical and capillary attraction and the attraction of cohesion: whether it be a modification of gravity, or that some new and unknown power comes into action, does not appear; but as a change in the law of the force takes place at one end of the scale, it is possible that gravitation may not remain the same throughout every part of space. Perhaps the day may come when even gravitation, no longer regarded as an ultimate principle, may be resolved into a yet more general cause, embracing every law that regulates the material world.

The action of the gravitating force is not impeded by the intervention even of the densest substances. If the attraction of the sun for the centre of the earth, and of the hemisphere diametrically opposite to him, were diminished by a difficulty in penetrating the interposed matter, the tides would be more obviously affected. Its attraction is the same also, whatever the substances of the celestial bodies may be; for if the action of the sun upon the earth differed by a millionth part from his action upon the moon, the difference would occasion a periodical variation in the moon's parallax whose maximum would be the $\frac{1}{15}$ of a second, and also a variation in her longitude amounting to several seconds, a supposition proved to be impossible, by the agreement of theory with observation. Thus all matter is pervious to gravitation, and is equally attracted by it.

As far as human knowledge extends, the intensity of gravitation has never varied within the limits of the solar system; nor does even analogy lead us to expect that it should; on the contrary, there is every reason to be assured that the great laws of the universe are immutable, like their Author. Not only the sun and planets, but the minutest particles, in all the varieties of their attractions and repulsions,—nay, even the imponderable matter of the electric, galvanic, or magnetic fluid,—are all obedient to permanent laws, though we may not be able in every case to resolve their phenomena into general principles. Nor can we suppose the structure of the globe alone to be exempt from the universal fiat, though ages may pass before the changes it has undergone, or that are now in progress, can be referred to existing causes with the same certainty with which the motions of the planets, and all their periodic and secular variations, are referable to the law of gravitation. The traces of extreme antiquity perpetually occurring to the geologist give that information as to the origin of things in vain looked for in the other parts of the universe. They date the beginning of time with regard to our system; since there is ground to believe that the formation of the earth was contemporaneous with that of the rest of the planets; but they show that creation is the work of Him with whom 'a thousand years are as one day, and one day as a thousand years.'

It thus appears that the theory of dynamics, founded upon terrestrial phenomena, is indispensable for acquiring a knowledge of the revolutions of the celestial bodies and their reciprocal influences. The motions of the satellites are affected by the forms of their primaries, and the figures of the planets themselves depend upon their rotations. The symmetry of their internal structure proves the stability of these rotatory motions, and the immutability of the length of the day, which furnishes an invariable standard of time; and the actual size of the terrestrial spheroid affords the means of ascertaining the dimensions of the solar system, and provides an invariable foundation for a system of weights and measures. The mutual attraction of the celestial bodies disturbs the fluids at their surfaces, whence the theory of the tides and the oscillations of the atmosphere. The density and elasticity of the air, varying with every alternation of temperature, lead to the consideration of barometrical changes, the measurement of heights, and capillary attraction; and the doctrine of sound, including the theory of music, is to be referred to the small undulations of the aërial medium. A knowledge of the action of matter upon light is requisite for tracing the curved path of its rays through the atmosphere, by which the true places of distant objects are determined, whether in the heavens or on the earth. By this we learn the nature and properties of the sunbeam, the mode of its propagation through the etherial fluid, or in the interior of material bodies, and the origin of colour. By the eclipses of Jupiter's satellites, the velocity of light is ascertained, and that velocity, in the aberration of the fixed stars, furnishes the only direct proof of the real motion of the earth. The effects of the invisible rays of light are immediately connected with chemical action; and heat, forming a part of the solar ray, so essential to animated and inanimated existence, whether considered as invisible light or as a distinct quality, is too important an agent in the economy of creation not to hold a principal place in the order of physical science. Whence follows its distribution over the surface of the globe, its power on the geological convulsions of our planet, its influence on the atmosphere and on climate, and its effects on vegetable and animal life, evinced in the localities of organized beings on the earth, in the waters, and in the air. The connexion of heat with electrical phenomena, and the electricity of the atmosphere, together with all its energetic effects, its identity with magnetism and the phenomena of terrestrial polarity, can only be understood from the theories of these invisible agents, and are probably principal causes of chemical affinities. Innumerable instances might be given in illustration of the immediate connexion of the physical sciences, most of which are united still more closely by the common bond of analysis which is daily extending its empire, and will ultimately embrace almost every subject in nature in its formulæ.

These formulæ, emblematic of Omniscience, condense into a few symbols the immutable laws of the universe. This mighty instrument of human power itself originates in the primitive constitution of the human mind, and rests upon a few fundamental axioms which have eternally existed in Him who implanted them in the breast of man when He created him after His own image.

EXPLANATION OF TERMS.

EXPLANATION OF TERMS.

Aberration. An apparent annual motion in the fixed stars, occasioned by the velocity of light combined with the real velocity of the earth in its orbit.

Absorbent media. Substances either solid, liquid, or fluid, which imbibe the rays of light or heat.

Accidental colours. If the eye has been dazzled by looking steadily at a bright colour, as, for example, at a red wafer, upon turning it to a white object a bluish-green image of the wafer will appear. Bluish-green is therefore the accidental colour of red, and *vice versâ.* Each tint has its accidental colour. When the real and accidental colours are of equal intensity, the one is said to be the complementary colour of the other, because the two taken together make white light.

Acceleration. A secular variation in the mean motion of the moon.

Aëriform. Having the form of air.

Aërolite. A meteoric stone.

Aërostatic expedition. Ascent in a balloon.

Affinity or *cohesive force.* The force with which the particles of bodies resist separation.

Algæ. Sea weeds or marine plants.

Aliquot parts. The parts into which a quantity is divided when no remainder is left.

Altitude. The height of an object above the horizon.

Analysis. Mathematical reasoning conducted by means of abstract symbols.

Analyzing plate. A piece of glass or a slice of a crystal used for examining the properties of polarized light.

Analytical formula or *expression.* A combination of symbols expressing a series of calculation, and including every particular case that can arise from a general law.

Angle of position of a double star. The angle which a line joining the two stars makes with one parallel to the meridian.

Angular velocity. The swiftness with which the particles of a revolving body move. It is proportional to the velocity of each particle divided by its distance from the axis or centre of rotation.

Annual equation. A periodical inequality in the motion of the moon going through its changes in a year.

Annual parallax. See *Parallax.*

Antimony. A metal.

Antennæ. The thread-like horns on the heads of insects.

Aphelion. The point in which a planet is at its greatest distance from the sun—the point A in *fig.* 8, S being the sun.

Apogee. The point in which the sun or moon is farthest from the earth.

Apparent motion. The motion of the celestial bodies as viewed from the earth.

Apparent diameter. See *Diameter.*

Apparent time. See *Time.*

Apsides. The extremities A and P, *fig.* 8, of the major axis of an orbit, or the points in which a planet is at its greatest and least distances from S the sun; also those in which a satellite is at its greatest and least distances from its planet.

Arc of the meridian. Part of a plane curve passing through the poles of the earth, and along its surface.

Areas. Superficial extent. In astronomy, they are the spaces passed over by the radius vector of a celestial body.

Arithmetical progression. A series of quantities or numbers continually increasing or diminishing by the same quantity; as, for example, the natural numbers 0, 1, 2, 3, 4, &c., which continually increase by one.

Armature. A piece of soft iron connecting the poles of a horse-shoe magnet.

Astronomical or *solar day.* The time between two consecutive true noons or midnights.

Atmospheric refraction. See *Refraction.*

Aurora. A luminous appearance in the heavens, frequently seen in high northern and southern latitudes.

Axis of rotation. The line, real or imaginary, about which a body revolves. The imaginary line passing through both poles and the centre of the earth is the axis of the earth's rotation.

Axis of a prism. The line *a b*, *fig.* 11, passing through the centre of a prism parallel to its sides.

Axis of a telescope. An imaginary line passing through the centre of the tube.

Axis of an ellipse. See *Ellipse*, line A B, *fig.* 2.

Base. In surveying, a base is a line measured on the surface of the earth, and assumed as an origin from whence the angular and linear distances of remote objects may be determined.

Binary system of stars. Two stars revolving about each other.

Bissextile. Leap year, every fourth year.

Caloric. The material of heat; heat being the sensation.

Centre of gravity. A point in a body, which, if supported, the body will remain at rest.

Capillary attraction. The attraction of tubes with a very minute bore, such as thermometer tubes, which causes liquids to ascend and remain suspended within them.

Centrifugal force. The force with which a revolving body tends to fly from the centre of motion. The direction of this force is in the tangent to the path the body describes.

Circumference. The boundary of a circle.

Civil day. The time comprised between two consecutive returns of the sun to the same meridian.

Civil or *tropical year.* The time comprised between two consecutive returns of the sun to the same solstice or equinox.

Chemical rays. The rays of the solar spectrum which do not produce light but destroy vegetable colours.

Chronometer. A watch which measures time more accurately than those in common use.

Coal measures. The strata which contain beds of coal.

Cobalt. A metal.

Cohesion. The force with which the parts of bodies resist any endeavour to separate them. Hardness, softness, tenacity, fluidity, and ductility, are modifications of cohesion.

Collecting wires, or *Collectors.* Wires for collecting and conveying electricity.

Complementary colours. See *Accidental colours.*

Compression of a spheroid. Flattening at the poles. It is equal to the difference between the greatest and least diameters divided by the greatest.

Concave mirror. A polished curved surface which, being hollow, reflects parallel rays of light so as to make them tend to meet.

Concentric. Having the same centre.

Conductor. A substance which conducts the electric fluid.

Fig. 1.

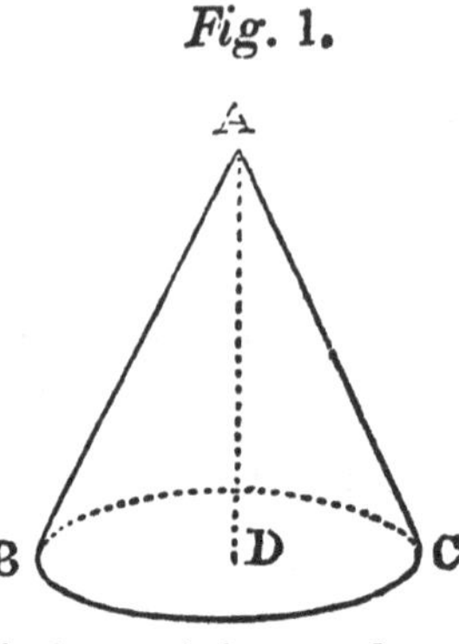

Cone. A solid figure A B C, like a sugar-loaf, of which A is the apex, A D the axis, and the plane B C the base. The axis may or may not be at right angles to the base, and the base may be a circle, an ellipse, or any other line. When the axis is at right angles to the base, the figure is a right cone.

Conic sections. Lines formed by a plane cutting a cone, of which there are five. If a right cone with a circular base, be cut at right angles to

Fig. 2.

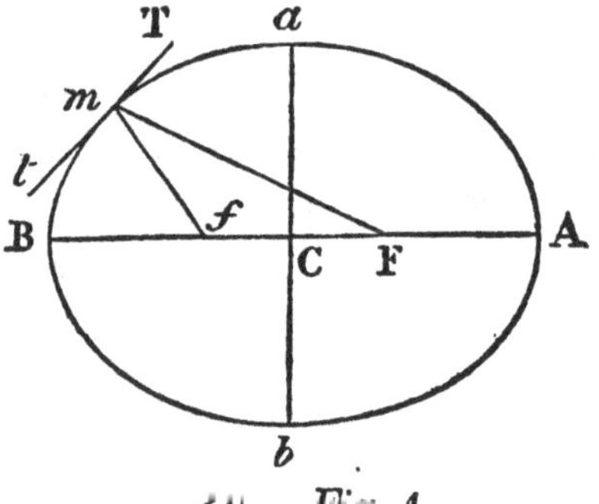

Fig. 3.

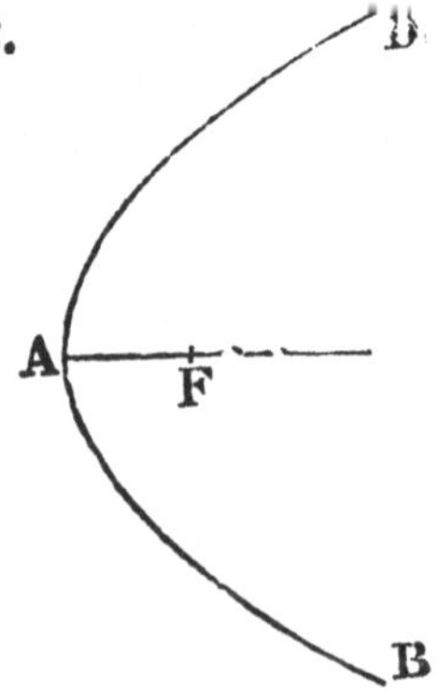

Fig. 4.

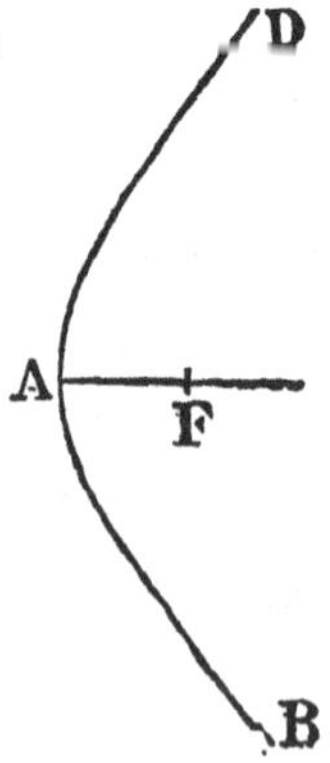

the base by a plane passing through the apex, the section will be a triangle. If the cone be cut through both sides by a plane parallel to the base, the section will be a circle. If the cone be cut slanting quite through both sides, the section will be an ellipse, B *a* A *b*, *fig*. 2. If the cone be cut parallel to one of its sloping sides, the section will be a parabola, D A B, *fig*. 3 ; and if the section cut only one side of the cone, and be not parallel to the other, it will be a hyperbola, D A B, *fig*. 4.

Configuration. The position of bodies with regard to one another.

Conjunction. A planet is said to be in conjunction when it has the same longitude with the sun.

Constellations. Groups of stars to which the names of men and animals have anciently been given. The whole starry firmament is divided into such groups.

Contrasted colours. See *Accidental colours.*

Converging. Tending to the same point.

Convex mirror. A polished curved surface which, being protuberant, reflects parallel rays of light, so as to make them diverge.

Cosine of an arc or angle. In *fig*. 5, A D is the co-sine of the arc C B, and of the angle B A C.

Fig. 5.

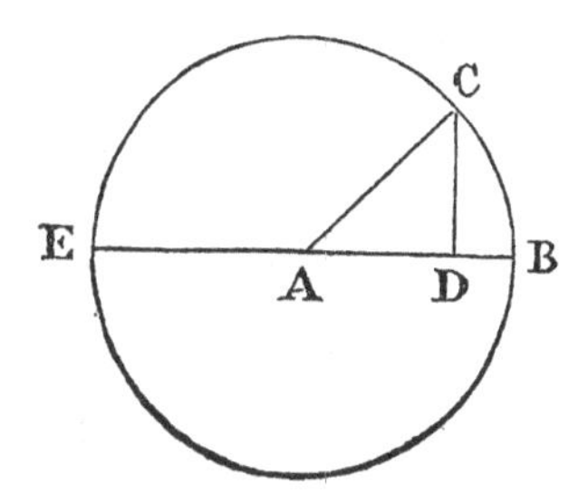

Crystal. A chemical or mineral substance having a regular form.

Curves of double curvature. Lines curved in two directions, so that no two of their indefinitely small parts lie in the same plane, as a corkscrew or a curved line drawn obliquely on the side of a cylinder which has its own curvature, at the same time that it partakes of the curvature of the surface on which it is drawn.

Curves of the second order. The conic sections. In a circle, the relation of the part A D, *fig.* 5, of the diameter to the perpendicular D C, is the same for every point in the circumference. The two lines A D, D C are called co-ordinates. The relation of these co-ordinates to one another is different in different curves, but remains invariable in any one curve; and lines are said to be of the first or second order, according as this relation can be expressed by the simple lines themselves, or by their squares and products.

Fig. 6.

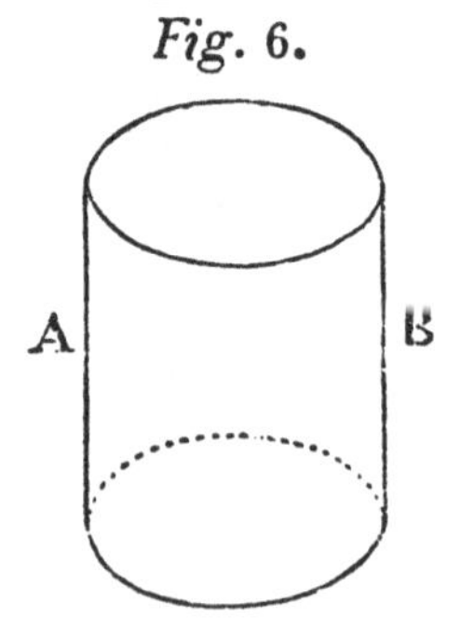

Cylinder. A solid A B, *fig.* 6, formed by the revolution of a parallelogram about one of its sides.

Declination. The angular distance of a celestial object from the celestial equator.

Density. The quantity of matter in a given bulk.

Diagonal. A line drawn from angle to angle of a four sided figure, as C D, *fig.* 10.

Diameter. A straight line, E B, *fig.* 5, passing through the centre, and terminated both ways, by the sides or surface of a figure.

Diameter, apparent. The diameter of a body as seen from the earth.

Diaphanous. Transparent.

Dicotyledonous plants. Such as have seeds containing two lobes.

Dip. The angle formed by the direction of the magnetic force of the earth with the plumb-line.

Dipping needle. An instrument for measuring the dip of a magnetized needle.

Direct motion of a celestial body. Motion from west to east, according to the order of the signs of the zodiac.

Disc. The apparent surface of a heavenly body.

Disintegration. Mouldering down, separating into parts.

Distance, mean. See *Mean distance.*

Distance, true. See *True distance.*

Distance, Perihelion. See *Perihelion distance.*

Diverging. Tending from a point.

Double refraction. The power which some substances possess of refracting or transmitting a ray of light in two pencils instead of one. If S I, *fig.* 13, be a ray of light falling upon a doubly refracting surface G g, it will be transmitted in two pencils I O, I E, so that the luminous point S will appear double, if viewed through the substance G *g*; whereas if G *g* were of glass or water, the ray S I would be transmitted in a single pencil, I O, and only one image of S would be seen.

Dynamics. The science of force and motion.

Ecliptic. The great circle traced in the starry heavens by the plane of the ecliptic.

Ecliptic, plane of. An imaginary plane passing through the earth s orbit, and extending to the starry heavens.

Elasticity. The property bodies possess of resuming their original form, when pressure is removed.

Elastic media. Atmospheric air, gas, ether, &c., which are highly compressible, and instantly resume their volume or bulk, when pressure is removed.

Electrics. Substances in which electricity may be excited, but which are incapable of conducting it.

Electric induction. The effect of electrified bodies to produce an electric state opposite to their own, in all bodies near them capable of receiving it.

Electro-magnetism. The science which determines the reciprocal action of electricity and magnetism.

Electro-magnets. Cylinders which have all the properties of magnets when a stream of electricity is passing through them.

Electro-dynamics. The science of the motion and reciprocal action of electric currents.

Electro-dynamic-cylinder. A hollow coil of copper wire, (*fig.* 7,) in the form of a corkscrew, the

Fig. 7.

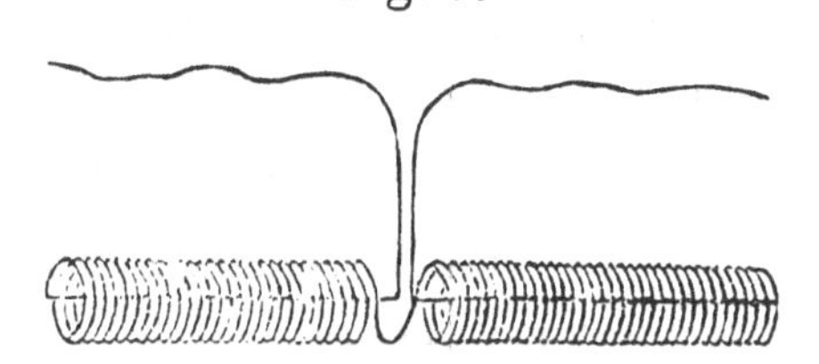

extreme parts of the wires of which are passed back through the centre of the coil, and being bent at right angles are brought out through its middle. There are several forms of this instrument, all of which have the same properties as magnets, when a galvanic current is passing through them.

Elements of an orbit. In an elliptical orbit there are six elements. Let P *n* A N (*fig.* 8) be the orbit of a planet, S the Sun, C N E *n* the plane of the ecliptic, and ♈ the first point of Aries, then the six

Fig. 8.

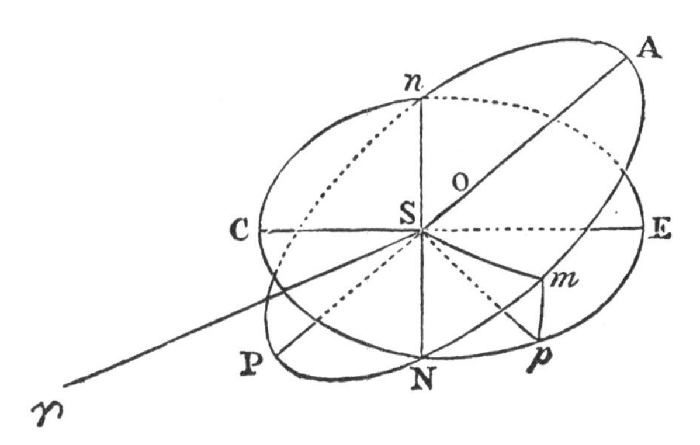

elements are the major axis P A, the excentricity S O, the longitude ♈ S P of P the perihelion, the longitude ♈ S N of N, the ascending node, the inclination of the orbit *n* A N on the plane of the ecliptic *n* E N, and ♈ S *m* the longitude of the body *m*, at any given instant called the longitude of the epoch. In a parabolic orbit, there are only five elements, since the major axis is infinite.

Ellipse. One of the conic sections. An ellipse may be drawn, by fixing the ends of a string to two points, F and *f* (*fig.* 2.) in a sheet of paper, and then carrying the point of a pencil round in the loop of the string kept stretched, the length of the string being greater than the distance between the two points. The points F and *f* are called the foci, and the distance F C is the excentricity, C being the centre of the ellipse; it is evident that the less F C is, the nearer does the ellipse approach the form of a circle. A B is the major axis, *a b* the minor axis, and F A the focal distance. From the construction, the length of the string, F *m f*, is equal to the major axis. If T *t* be a tangent to any point *m*, and F *m*, *f m* lines from the foci, the angle F *m* T is equal to the angle *f m t*; and as this is true for every point of the ellipse, it follows that, in an elliptical reflecting surface, rays of light or sound coming from the focus F will be reflected by the surface to the other focus *f*, since the angle of incidence is equal to the angle of reflection, by the theory of optics and acoustics.

Ellipsoid of revolution. A solid formed by the revolution of an ellipse about its axis. If the ellipse revolves about its minor axis, the ellipsoid will be oblate, or flattened at the poles, like an orange : if the revolution be about the major axis, the ellipsoid will be drawn out at the poles, or prolate, like an egg.

Ellipticity. Excentricity, or deviation from the circular or spherical form.

Elongation. The angular distance of a celestial body from the sun, as it would be seen from the centre of the earth.

Epoch. The assumed instant from whence all the subsequent and antecedent periods of a celestial body are estimated.

Equation of time. The difference between the time shown by a watch, and that given by a dial, or the difference of mean and true time.

Equation of the centre. The difference between the true and mean motion of a planet or satellite. At its maximum, it is equal to the excentricity of the orbit, since it is the difference of the motion of a body in an ellipse, and in a circle whose diameter is equal to the major axis of the ellipse.

Equator. The terrestrial equator is the equinoctial line. The celestial equator is the great circle traced in the starry heavens by the imaginary extension of the plane of the terrestrial equator.

Equinoxes. The vernal and autumnal equinoxes are two points in the heavens diametrically opposite to one another, that is 180° apart. The line in which the planes of the equator and ecliptic intersect passes through them. The vernal equinox is the point from whence the longitudes or angular distances of the celestial bodies are estimated; it is generally called the first point of Aries, though these two points have not coincided since the early ages of astronomy, about two thousand two hundred and thirty-two years ago, on account of the precession or retrograde motion of the equinoctial points.

Etherial medium. The ether or highly elastic fluid with which space is filled.

Evection. A certain periodic inequality in the motion of the moon.

Excentricity. The distances between the centre and focus of an ellipse, or C F, (*fig.* 2.)

Extraordinary refraction. See *Refraction.*

Extraordinary ray. See *Refraction.*

Focus. A point where converging rays or lines meet.

Focal distance. The line F A in the conic sections, (*fig.* 2, 3, and 4.)

Foci of an ellipse. Two points F and *f* (*fig.* 2.) in the major axis, such, that the sum of the two lines drawn from them to any point *m* in the ellipse is equal to the major axis A B.

Fossils, organic. The remains of ancient animals and plants embodied in the strata of the earth.

Fundamental note. The natural note of any sonorous body, as of a string or organ-pipe.

Galvanism. Electricity perpetually in motion, and produced by chemical action.

Galvanic battery. An instrument for producing galvanic electricity, constructed of alternate layers of two metals and a fluid.

Galvanic circuit. Three substances in contact, generating a stream of electricity, which flows in a perpetual circuit through them.

Galvanometer. An instrument for measuring the intensity of the galvanic force.

Genera of plants. The divisions of plants into families, each of which contains a variety of species.

General analytical expression. The representation in symbols of a series of reasoning, including every particular case of the subject in question.

Geometrical progression. A series of quantities increasing or diminishing by a continual multiplication or division by the same quantity, as the numbers 1, 2, 4, 8, 16, &c., which are constantly multiplied by 2, or the series 1, $\frac{1}{2}$, $\frac{1}{4}$, $\frac{1}{8}$, &c., which decreases by the continual division by 2.

Graphical construction of an orbit. The drawing of an orbit by ruler and compass from given observations.

Gravity. The attraction of matter, weight.

Gravitating force. The force with which matter attracts; its intensity varies inversely as the square of the distance; that is, the weight of a body decreases in proportion as the square of its distance from the centre of the earth increases, and *vice versâ.* But if the body be within the surface of the earth, the force varies inversely as the distance from the centre.

Gravitation. Sensible gravity.

Grylli. Grasshoppers, crickets, locusts, &c.

Harmonics. The doctrine of musical sound.

Harmonic sounds. The sympathetic notes heard along with the principal note of a musical string, or other sonorous body.

Harmonic divisions. The parts into which a vibrating musical string spontaneously divides itself, each of which gives a distinct note, besides the principal note arising from the vibration of the whole string.

Harmonic colours. Tints which become visible upon looking steadfastly at a bright coloured light, supposed to be analogous to the sympathetic notes of a musical string.

Helix. A curve like a corkscrew, whose turnings may either be circular or elliptical.

Homogeneous light. Rays of the same colour.

Homogeneous spheroid. A spheriod of uniform density.

Horizontal plane. An imaginary plane touching the surface of the earth in one point, and terminating on all sides in the horizon.

Horoscope. The relative positions of the planets at the time of a person's birth.

Hyperbola. One of the conic sections. An open curve, having two infinite branches, A B, A D, (*fig.* 4.) and a focus F, to which every point in the curve has a fixed relation.

Hypothesis. A system upon supposition. An assumption.

Iceland spar. A transparent and colourless carbonate of lime, consisting of fifty-six parts of lime, and forty-four of carbonic acid. It splits or cleaves into solids called rhombs (*fig.* 14.) which are bounded by six similar surfaces, whose sides are parallel, but the angles are not right angles: it possesses the property of double refraction in an eminent degree.

Impetus. A force whose intensity is measured by the mass of a body and the square of its velocity conjointly.

Incidence of light. The angle which rays make with a perpendicular to the surface upon which they fall. Let A B, *fig.* 9, be the reflecting surface,

Fig. 9.

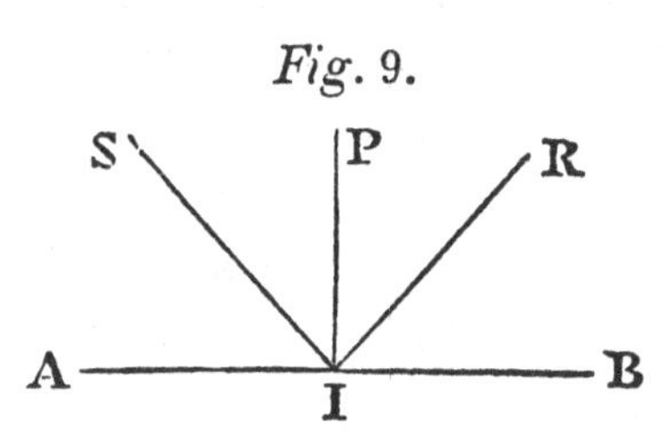

then S I is the incident and I R the refracted ray, the angle S I P being equal to the angle R I P.

Inclination of an orbit. The angle in *fig.* 8. which the plane of an orbit P *n* A N makes with the plane of the ecliptic C N E *n*.

Indigenous. Native to a particular spot or country.

Inertia. The disposition of matter to remain in its state of rest or motion.

Interference of undulations. The combination of two series of waves in a fluid so as to augment, diminish, or destroy each other.

Isochronous. In equal times.

Isothermal lines. Imaginary lines passing through such places as have the same mean annual temperature.

Isogeothermal lines. Imaginary lines passing through all those places within the surface of the earth, where the mean internal temperature is the same.

Kepler's laws. Three laws in the planetary motions discovered by Kepler, which furnish the data from whence the principle of gravitation is established: they are, First, that the radii vectores of the planets and comets describe areas proportional to the time: Second, that the orbits of the planets and comets are conic sections, having the sun in one of their foci; and third, that the squares of the periodic times of the planets are proportional to the cubes of their mean distances from the sun. These laws extend also to the satellites.

Latent heat. Caloric existing in all bodies, which is not sensible, and cannot be detected by the thermometer.

Latitude. Terrestrial latitude is the angular distance between the vertical or plumb-line at any place and the plane of the equator. Celestial latitude is the height of a heavenly body above or below the plane of the ecliptic, as *m* s *p*, *fig.* 8; when above, it has north, and when below that plane, it has south latitude.

Length of a wave. The distance between two particles of an undulating fluid similarly displaced and moving similarly, consequently the length is the distance between two consecutive hollows or elevations.

Lens. A transparent substance with curved surfaces. The glasses of a telescope and of spectacles are lenses. A lens may be convex on both sides, or it may have both sides concave; one side may be convex and the other concave; one side plane and the other convex; or lastly, one side may be plane and the other concave.

Libration. A balancing motion.

Lines of the second order. The circle, ellipse, parabola, hyperbola, and generally such as are expressed algebraically by a quadratic equation. See *Curves of the second order.*

Lines of no variation. Imaginary lines passing through all places where the needle of the mariner's compass points to the true north, that is, to the pole of the earth's rotation.

Lines of perpetual snow. Imaginary lines passing through the limits of perpetual snow from the equator to the poles.

Longitude. Terrestrial longitude is the angular distance of a place from a meridian arbitrarily chosen, such as that of Greenwich.

Longitude of a heavenly body. The true longitude of a planet, as of *m*, *fig*. 8. is its angular distance ♈ s *m* from ♈ the vernal equinox, estimated on its elliptical orbit; its mean longitude is its angular distance from the same point, supposing the planet to move equably in a circle whose radius is equal to the mean distance of the body from the sun. The difference between the two is the equation of the centre.

Longitude of the perihelion. The angular distance of the perihelion of an orbit from the vernal equinox, as ♈ s P, *fig*. 8.

Longitude of the node. The angular distance of the node of an orbit from the vernal equinox as, ♈ s N, *fig*. 8.

Longitude of the epoch. The angular distance of a celestial body from the vernal equinox at the instant assumed as the origin of time whence all its subsequent and antecedent longitudes are estimated.

Lunar distance. The angular distance of the centre of a celestial object from the centre of the moon.

Magnetic equator. The imaginary line passing through those places where there is no dip, that is, where the compass needle is horizontal. It encircles the earth, but does not coincide with the terrestrial equator.

Magnetic meridian. The vertical plane passing through the direction of the needle of the compass at any place.

Magnetic poles. Points of the earth where the intensity of the magnetic force is greatest.

Magnetic induction. The effect of magnets to excite magnetism in bodies near them.

Magneto-electric induction. The effect of galvanic currents to produce magnetism in bodies near them capable of receiving it.

Major axis or *greatest diameter of an ellipse.* See *Ellipse*, A B, *fig.* 2.

Mass. The quantity of matter in a body. It is proportional to the density and volume conjointly.

Mathematics. The science of number and quantity.

Mean distance. The mean distance of a planet from the sun, or of a satellite from its planet, is equal to half the major axis of its orbit.

Mean longitude. See *Longitude.*

Mean motion. Equable motion in a circular orbit at the mean distance during the same time that the body accomplishes a revolution in its elliptical orbit.

Mean time. The time shown by clocks and watches well regulated.

Mechanics. The science of the equilibrium and motion of bodies.

Meridian. A vertical plane passing through the poles of the earth.

Meteorites. Stones which fall from the heavens.

Mica. A certain mineral.

Minor axis. See *Ellipse.*

Minus. Less. The sign of Subtraction.

Molecules. The indefinitely small or ultimate particles of matter.

Momentum. Force measured by the mass and simple velocity conjointly.

Monocotyledonous plants. Such as have seeds of one lobe.

Moon's southing. The time when the moon comes to the meridian of any place, which happens about forty-eight minutes later each day.

Multiple systems of stars. Three or more stars revolving about their common centre of gravity.

Nebulæ White misty appearance in the heavens like the milky way; some of them, when viewed with powerful telescopes, are found to be clusters of stars, others always retain the cloudy form.

Nebulosity of comets. The coma or misty appearance which always surrounds their heads, and of which their whole mass is often composed.

Nickel. A metal.

Nodes. The two opposite points N and *n*, *fig.* 8, in

which the orbit N A n P of a planet or comet intersects the plane C N E n of the ecliptic. Part, N A n, of the orbit lies above the plane of the ecliptic, and part, n P N, below it. The ascending node N is the point through which the body passes in rising above the plane of the ecliptic, and the descending node n is the point in which the body sinks below it. The nodes of a satellite's orbit are the points in which it intersects the plane of the orbit of its primary.

Nodes, line of. The intersection N n, *fig.* 8. of the plane of the orbit of a planet or comet with the plane of the ecliptic. It passes through S, the centre of the sun.

Nodal points. Points of a sonorous body which remain at rest during its vibrations.

Nodal lines. Lines of sonorous surfaces which remain at rest during their vibrations.

Non-electrics. Substances in which electricity cannot be sensibly excited by friction.

Nucleus of a comet. The part of its head which appears to be dense. Frequently they have none.

Nucleus of the earth. The solid part.

Nutation. A variation in the obliquity of the ecliptic from the attraction of the sun and moon on the protuberant matter at the terrestrial equator.

Nutation of the lunar orbit. A variation in the inclination of the lunar orbit from the action of the matter at the earth's equator on the moon. It is the reaction of terrestrial nutation.

Oasis. A fertile spot in a desert.

Oblate spheroid. A solid like an orange, which may be formed by the rotation of an ellipse about its minor axis, and is therefore flattened at the poles.

Obliquity of the ecliptic. The angle formed by the plane of the terrestrial equator with the plane of the ecliptic.

Oscillation. A motion to and fro, like the pendulum of a clock.

Occultation. The eclipse of a star or planet by the moon or by another planet.

Opposition. A body is said to be in opposition when its longitude differs from that of the sun by 180°.

Optics. The science of light and colours.

Optic axis of a crystal. A ray of light passing through a doubly refracting crystal, such as Iceland spar, is generally divided into two rays, but in certain directions it is transmitted in one ray only: these directions are called the optic axes of a crystal.

Orbit. The track or path of a celestial body in the heavens.

Ordinary refraction. See *Refraction.*

Ordinary ray. See *Refraction.*

Parabola. One of the conic sections. It is the line described by a cannon ball, and has two infinite branches, A B, A D, *fig.* 3. and there is a point F within it called the focus, to which every point in the curve bears a certain relation.

Parabolic elements. See *Elements of an orbit.*

Parallax. The angle under which we view an object; it therefore diminishes as the distance increases.

Parallax of a celestial object. The angle which the radius of the earth would be seen under, if viewed from that object.

Parallax, horizontal. The parallax of a celestial body when in the horizon. Parallax is then at its maximum; it decreases as the height of the body above the horizon increases.

Parallax, annual. The angle which the diameter of the earth's orbit would be seen under, if viewed from a celestial body, as a fixed star.

Parallactic motion. The motion of a body is said to be parallactic when the space described by it subtends or is seen under a sensible angle.

Parallelogram. A four-sided plane figure, A B, *fig.* 10. whose opposite sides are parallel: the dia-

Fig. 10.

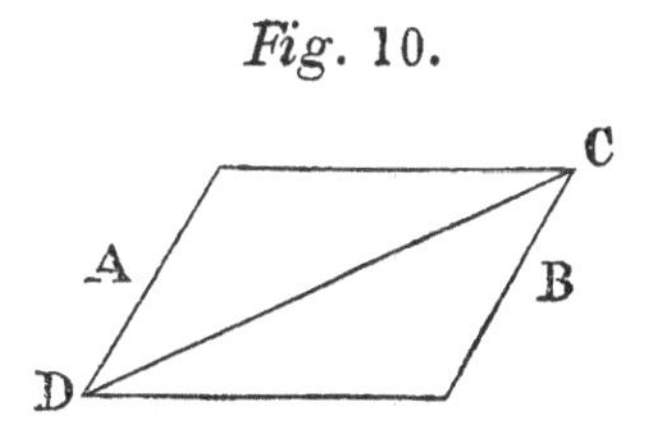

meter is the straight line joining two of its opposite angles.

Passage at the perihelion. The passage of a body through the point of its orbit that is nearest to the sun.

Penumbra. The shadow or imperfect darkness which precedes and follows an eclipse.

Perigee. The points in which the sun and moon are nearest to the earth.

Perihelion. The point P *fig.* 8. of an orbit which is nearest to the sun.

Perihelion distance. The shortest distance of a planet or comet from the sun, P S, *fig.* 8.

Periodic inequality. An irregularity in the motion of a celestial body requiring a comparatively short time for its accomplishment.

Periodic time. The time in which a planet or comet performs a revolution round the sun, or a satellite about its primary.

Perturbations. Irregularities in the motions of bodies from some disturbing cause.

Phanerogamous plants. Such as have apparent flowers and seeds.

Phases of the moon. The periodic changes in the enlightened part of her disc from a crescent to a circle, depending upon her position with regard to the sun and earth.

Phases of an undulation. Alternate changes in the surface or density of a fluid. The fluid particles in the tops or in the hollows of a series of waves are in the same phases, because their displacement and motion are equal and in the same direction;

whereas the fluid particles in the tops of a series of waves are in different phases from those in the hollows, because the displacement and motion of the first are equal, but opposite to those of the second. For example: in waves of water, the particles in the tops have arrived at their greatest elevation, and are beginning to sink down, whereas those in the hollows have reached their greatest depression, and are beginning to rise up.

Phenomena. Appearances.

Physical. Belonging to material nature.

Physico-mathematical sciences. Sciences in which natural phenomena are explained by mathematical reasoning.

Pitch in music. The depth or shrillness of a note. It depends upon the number of vibrations the sonorous body makes in a second. The more rapid the vibration the higher the pitch.

Plane. Length and breadth without thickness.

Plane of reflection. The plane passing through the incident and reflected rays of light or sound as S I, I R, *fig.* 9. It is perpendicular to the reflecting surface.

Plane of refraction. The plane passing through the incident and refracted rays of light S I and I O, *fig.* 13. It is perpendicular to the refracting surface.

Plane of polarization. The plane passing through the incident and polarized ray. It is at right angles to the plane of reflection, but deviates from the plane of ordinary refraction.

Plus. More; the sign of addition.

Polarity. The tendency of magnetized bodies to point to the magnetic poles of the earth.

Polarized light. Light which by reflection or refraction at a certain angle, or by refraction in certain crystals, has acquired the property of exhibiting opposite effects in planes at right angles to each other. This property is explained on the undulatory theory by supposing the particles of the ether to vibrate in one plane.

Polarization, circular. The property which light acquires, by transmission through quartz and certain liquids, of producing a succession of appearances which follow each other in a circular order, as the thickness of the medium is increased. This property is explained on the undulatory theory by supposing the particles of the ether to vibrate in circles one after the other, the undulation going on in a circular helix like a cork-screw penetrating a cork.

Polarization, elliptical. The property which light acquires, by reflection at the surfaces of metals and in other ways, of producing appearances partly analogous to those of circular polarization. It is explained by supposing the undulation to follow the course of an elliptical helix.

Poles. The extremities of the axis about which a body revolves.

Poles of the earth. The extremities of the axis of diurnal rotation.

Poles, magnetic. Points in the earth where the intensity of the magnetic force is a maximum. Of these there are certainly three, probably four, all of which differ from the poles of rotation.

Poles of a magnet. Points in a magnet where the intensity of the magnetic force is a maximum; one of these attracts and another repels the same pole of another magnet.

Poles of maximum cold. Points in the surface of the earth where the mean annual temperature is a maximum. There are several, but none of them coincide with the poles of rotation.

Precession of the equinoxes. A retrograde motion of the equinoctial points in consequence of the action of the sun and moon upon the protuberant matter at the earth's equator.

Primary. In astronomy signifies the planet about which a satellite revolves.

Prism. A triangularly or polygonally shaped piece of glass or other substance, like a three or more cornered stick, as *fig.* 11.

Fig. 11.

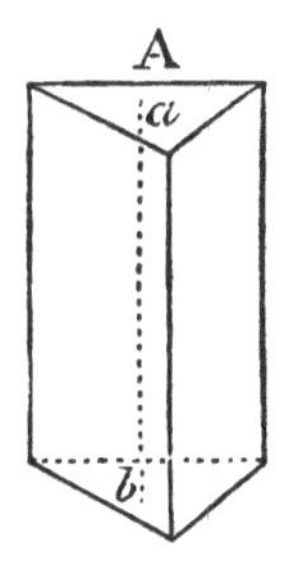

Prism, a doubly refracting. A prism made of a doubly refracting substance, as Iceland spar.

Prismatic colours. The colours of the rainbow.

Projected. Thrown; transferred by means of lines.

Projection. A line or surface is said to be projected upon a plane when parallel straight lines are drawn from every point of them to the plane. The projection of an orbit is therefore its day-light shadow, since the sun's rays are sensibly parallel.

Prolate spheroid. A solid figure something like an egg. See *Ellipsoid.*

Pulse. A vibration.

Pyramid. A solid bounded by a base having several sides, and by a number of triangular planes whose summits meet in one point called the apex, as *fig.* 12.

Fig. 12.

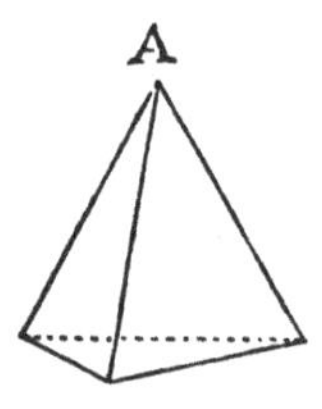

Pyrometer. An instrument for measuring intense degrees of heat.

Quadrant. Ninety degrees, the fourth part of a circle.

Quadrature. A celestial body is said to be in quadrature when it is ninety degrees distant from the sun.

Quartz. Rock crystal; a siliceous mineral whose primitive form is a rhomboid, *fig.* 14, but it is generally crystallized in six-sided prisms terminated by six-sided pyramids.

Radiation. An emission of rays.

Radius, equatorial. A line drawn from the centre of a spheroid to its equator.

Radius, polar. A line drawn from the centre of a spheroid to its pole.

Radius of a sphere. Any straight line drawn from the centre of a sphere to its circumference.

Radius vector. The imaginary line joining the centre of the sun and the centre of a planet or comet, or the centre of a planet and that of its satellite, as S *m*, *fig.* 8.

Ratio. A fraction expressing the relation which one quantity bears to another. Proportion is the equality of ratios.

Rectangle. A four-sided plane figure, in which all the angles are right angles, and its opposite sides equal and parallel. When all the sides are equal, it is a square.

Reflection. The bending back of rays of light or sound from a surface. The angles made by the rays with a perpendicular to the surface, in coming and going, are equal. If the ray, S I, (*fig.* 9) be reflected by a surface A B, in the direction I R, then the angle S I P is equal to R I P.

Refraction. The bending or breaking of a ray of light in passing through media of different densities, as in going from air into water or glass, and the contrary. If G *g* (*fig.* 13.) be a re-

Fig. 13.

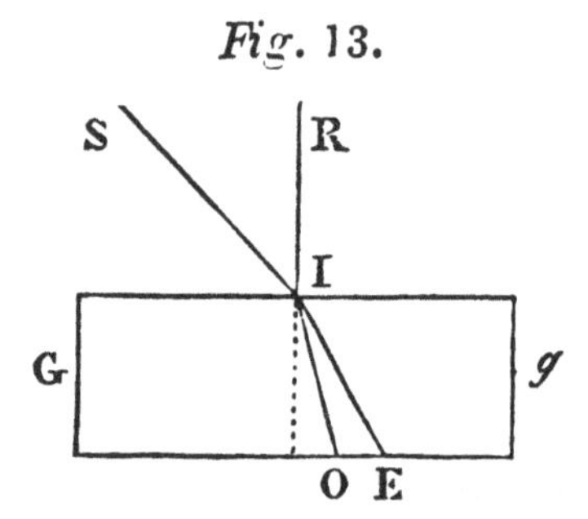

fracting medium, as a piece of glass, then S I is the incident, and I O the refracting ray.

Refraction, ordinary. Light is said to suffer ordinary refraction, when both the incident and refracted rays are in a plane at right angles to the refracting surface. This plane is called the plane of ordinary refraction, and the refracted ray is named the ordinary ray.

Refraction, extraordinary. Light is said to suffer extraordinary refraction, when it is refracted in a different plane from that of ordinary refraction. The plane in question is called the plane of extraordinary refraction, and the ray so refracted is named the extraordinary ray. In Iceland spar, and other doubly refracting substances, with one optic axis, the incident ray is split into two, one of which suffers ordinary, and the other extraordinary refraction, but in all doubly refracting substances, having two optic axes, both rays suffer extraordinary refraction.

Resulting force. The force resulting from the joint effects of a number of forces.

Retrograde motion of a celestial body. Its motion from east to west, or contrary to the signs of the zodiac.

Revolution of a planet. Its motion round the sun.

Revolution, sidereal. The consecutive returns of a planet to the same star.

Revolution, tropical. The consecutive returns of a planet to the same tropic or equinox.

Rhomb. A plane four-sided figure, whose opposite sides are equal and parallel, but all its sides are not equal, nor are its angles right angles.

Rhomboid or rhombohedron. A solid formed by six planes; the opposite planes being equal and similar rhombs parallel to one another, but all the planes are not necessarily equal nor similar, nor are its angles right angles (*Fig.* 14.)

Fig. 14.

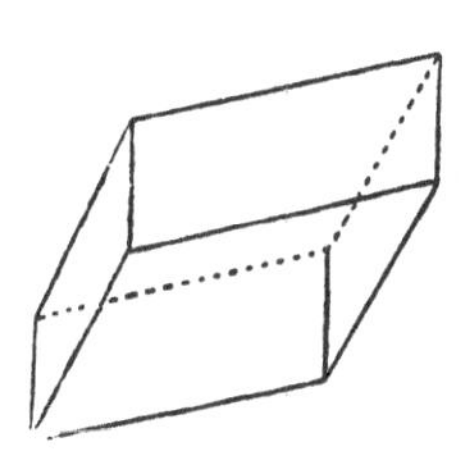

Rotation. The motion of a body round an axis.

Sauri or Saurians. Reptiles of the lizard kind, as crocodiles.

Secular inequalities. Variations in the motions of

the heavenly bodies, requiring many ages for their accomplishment.

Sidereal day. The time included between two consecutive transits of the same star at the same meridian.

Sidereal year. The time included between two consecutive returns of the sun to the same star.

Sine. The perpendicular drawn from the extremity of an arc to the diameter of a circle, C D, (*fig.* 5,) is the sine of the arc C B.

Solstices. The points in which the sun is farthest from the equator.

Solar spectrum. The coloured image of the sun refracted through a prism.

Space. The boundless region which contains all creation.

Species of plants. Plants of the same kind.

Sphere. A solid formed by the rotation of a semicircle about its diameter.

Spheroid of revolution, or *Ellipsoid.* A solid formed by the revolution of an ellipse about one of its axes. The spheroid will be oblate or prolate, according as the revolution is performed about the minor or major axis of the ellipse. Spheroids are sometimes irregular in their form.

Spiral. A curve like a watch spring. It may be circular, like a thread wound about a round rod; or elliptical, like a thread winding about an oval stick.

Stratum. A layer.

Subtend. To be opposite. In *fig.* 5, the arc C B subtends the angle C A B.

Sulphate of lime. A mineral capable of being split into thin transparent plates: it consists of 32·7 of lime, 46·3 of sulphuric acid, and 21 of water.

Synodic revolution of the moon. The time between two consecutive new or full moons.

Syzygies. The points in the moon's orbit where she is new or full.

Tangent. A straight line touching a curve in one point, as T *t* in *fig.* 2.

Tangential force. A force in the direction of the tangent.

Time, true. Time shown by a dial, or apparent time.

Time, mean. Time shown by ordinary clocks and watches.

Thermo-electric currents. Streams of electricity, excited by heat.

Transit. The passage of a body across the meridian of a place.

Transit of Venus and Mercury. The apparent passage of these planets across the sun's disc.

Trigonometrical measurements. Mensuration of the surface of the earth by a series of triangles.

Tropical year. The period between the consecutive returns of the sun to the same tropic or solstice.

True distance. The actual distance of a body from the sun, or of a satellite from its planet.

Undulation. A wave.

Undulatory theory. The mechanical principles of the motion of waves.

Vapour. Steam.

Variation. A periodic inequality in the motion of the moon.

Variation of the compass. The deviation of the compass needle from the true north.

Vertical. The direction of the plumb-line.

Vertical plane. A plane passing through the plumb-line, consequently at right angles to the horizon.

Vesicles. Small hollow spheres of water.

Vibration. A motion to and fro.

Visual ray. A ray of light coming from any object to the eye.

Volta-electric induction. The disposition of electric currents to produce similar currents in bodies near them capable of receiving them.

INDEX.

THE END.

Printed in the United States
By Bookmasters